Juniata College
Beeghly Library
Aaron and Clara
Rabinowitz
Memorial Library Fund

The Calculus of Selfishness

PRINCETON SERIES IN THEORETICAL AND COMPUTATIONAL BIOLOGY

Series Editor, Simon A. Levin

The Calculus of Selfishness,
by Karl Sigmund

The Geographic Spread of Infectious Diseases: Models and Applications,
by Lisa Sattenspiel with contributions from Alun Lloyd

Theories of Population Variation in Genes and Genomes,
by Freddy Bugge Christiansen

Analysis of Evolutionary Processes,
by Fabio Dercole and Sergio Rinaldi

Mathematics in Population Biology,
by Horst R. Thieme

Individual-based Modeling and Ecology,
by Volker Grimm and Steven F. Railsback

The Calculus of Selfishness

Karl Sigmund

PRINCETON UNIVERSITY PRESS
PRINCETON AND OXFORD

Published by Princeton University Press, 41 William Street, Princeton, New Jersey 08540

In the United Kingdom: Princeton University Press, 6 Oxford Street, Woodstock,
Oxfordshire OX20 1TW

Library of Congress Cataloging-in-Publication Data

Sigmund, Karl, 1945-
The calculus of selfishness / Karl Sigmund.
p. cm. — (Princeton series in theoretical and computational biology)
Includes bibliographical references and index.
ISBN 978-0-691-14275-3 (hardcover : alk. paper) 1. Game theory. 2. Cooperativeness—Moral and ethical aspects. 3. Evolution (Biology)—Mathematics. I. Title.
HB144.S59 2009
306.3′4—dc22 2009015030

British Library Cataloging-in-Publication Data is available

This book has been composed in Times & Abadi

Printed on acid-free paper. ∞

press.princeton.edu

Printed in the United States of America

10 9 8 7 6 5 4 3 2 1

Contents

Preface

You need not be a scheming egotist to pick up *The Calculus of Selfishness*. It is enough to be interested in the logic behind the ceaseless give-and-take pervading our social lives. The readership I had in mind, when writing this book, consists mostly of undergraduates in economics, psychology, or evolutionary biology. But simple models of social dilemmas are of general interest.

As the word *Calculus* in the title gives away, you will need a modicum of elementary mathematics. Beyond this, all the game-theory expertise you need is painlessly provided step by step. As to the *Selfishness* in the title, I do not mean blind greed, of course, but "enlightened self-interest," by which, according to Tocqueville, "Americans are fond of explaining almost all the actions of their lives; . . . . They show with complacency how an enlightened regard for themselves constantly prompts them to assist each other." Such complacency may well be justified; but theoreticians cannot share it. Most of them feel that it is hard to understand why self-interested agents cooperate for their common good.

In the New Year 2000 edition of *Science*, the editors listed "The evolution of cooperation" as one of the ten most challenging problems of the century. My book certainly does not claim to solve the problem. Having worked for twenty years in the field, I know that it progresses far too fast to allow an encyclopedic presentation, even when restricted to cooperation in human societies, which is a tiny fraction of all the cooperation encountered in biology.

Rather than trying to address all aspects, this book concentrates on one issue only, the reciprocity between self-interested individuals, and explores it for a small number of elementary types of interactions. The method is based on an evolutionary approach: more successful strategies become more frequent in the population. We neglect family ties, or neighborhood relations, or individual differences, or cultural aspects. It is best to state this self-limitation right at the beginning. I hope not to convey the impression that family ties, neighborhood relations, or individual aspects, etc., play no role in the evolution of cooperation and that it all reduces to self-interest; just as theoretical physicists writing a treatise on gravity do not imply, thereby, that other forces in the universe can be ignored. This being said, the current trend in economic life seems to lead away from nepotism, parochialism, and face-to-face encounters, and toward interactions between strangers in a well-mixed world.

The introduction (an entire chapter without any formulas) describes some of the most basic social dilemmas. Thinkers throughout the ages have been fascinated by the topic of self-regarding vs. other-regarding behavior, but the use of formal models and experimental games is relatively recent. Ever since Robert Trivers introduced an evolutionary approach to reciprocity, the Prisoner's Dilemma game serves as a kind of model organism to help explore the issue. But other games, such as the

Ultimatum, are quickly catching up. The most gratifying aspect of this development is the close connection between theoretical and experimental progress.

The second chapter provides a self-contained introduction to evolutionary game theory, stressing deterministic dynamics and stochastic processes, but tying this up with central notions of classical game theory, such as Nash equilibria or risk-dominance.

The third chapter provides a detailed discussion of repeated interactions, such as the Prisoner's Dilemma or the Snowdrift game, which allow exploration of direct reciprocity between the same two players meeting again and again. In particular, simple strategies based on the outcome of the previous round (such as *Tit for Tat*) or implemented by finite automata (such as *Contrite Tit for Tat*) offer a wide range of behavior.

The fourth chapter is devoted to indirect reciprocity. Here, players interact at most once, but they are aware of the past behavior of their one-shot partner. This introduces topics such as moral judgment or concern for reputation. Strategies based on the assessment of interactions between third parties allow the emergence of types of cooperation immune to exploitation, because they are channeled towards cooperators only.

The fifth chapter deals with the Ultimatum and the Trust game. Such games allow one to tackle the issues of fairness and trust, and provide, as a kind of side benefit, a framework for analyzing the roles of positive and negative incentives. Again, reputation plays an essential role for cooperation to emerge.

The sixth chapter turns from interactions between two players to interactions within larger groups. In so-called Public Goods games, defection can be suppressed by rewards or sanctions. Such incentives, properly targeted, allow reciprocation in mixed groups of cooperators and defectors. An intriguing aspect concerns the role of voluntary, rather than compulsory, participation in the team effort. Coercion emerges more easily if participation is optional.

The short seventh chapter, finally, deals briefly with some of the many issues that were neglected, such as nepotism, localized interactions, or group selection.

Needless to say, this book owes much to my colleagues, many of whom read draft chapters and provided comments. In particular, I want to thank Christoph Hauert, Arne Traulsen, Hannelore De Silva (formerly Brandt), Hisashi Ohtsuki, Satoshi Uchida, Ulf Dieckmann, Tatsuya Sasaki, Simon Levin, Ross Cressman, Yoh Iwasa, Silvia De Monte, Christoph Pflügl, Christian Hilbe, Steve Frank, Simon Gächter, Benedikt Hermann, Dirk Semmann, Manfred Milinski, and Josef Hofbauer. Most of all, I am indebted to Martin Nowak, without whom this book could never have been written.

The Calculus of Selfishness

Chapter One

Introduction: Social Traps and Simple Games

1.1 THE SOCIAL ANIMAL

Aristotle classified humans as social animals, along with other species, such as ants and bees. Since then, countless authors have compared cities or states with bee hives and ant hills: for instance, Bernard de Mandeville, who published his *The Fable of the Bees* more than three hundred years ago.

Today, we know that the parallels between human communities and insect states do not reach very far. The amazing degree of cooperation found among social insects is essentially due to the strong family ties within ant hills or bee hives. Humans, by contrast, often collaborate with non-related partners.

Cooperation among close relatives is explained by *kin selection*. Genes for helping offspring are obviously favoring their own transmission. Genes for helping brothers and sisters can also favor their own transmission, not through direct descendants, but indirectly, through the siblings' descendants: indeed, close relatives are highly likely to also carry these genes. In a bee hive, all workers are sisters and the queen is their mother. It may happen that the queen had several mates, and then the average relatedness is reduced; the theory of kin selection has its share of complex and controversial issues. But family ties go a long way to explain collaboration.

The bee-hive can be viewed as a watered-down version of a multicellular organism. All the body cells of such an organism carry the same genes, but the body cells do not reproduce directly, any more than the sterile worker-bees do. The body cells collaborate to transmit copies of their genes through the germ cells—the eggs and sperm of their organism.

Viewing human societies as multi-cellular organisms working to one purpose is misleading. Most humans tend to reproduce themselves. Plenty of collaboration takes place between non-relatives. And while we certainly have been selected for living in groups (our ancestors may have done so for thirty million years), our actions are not as coordinated as those of liver cells, nor as hard-wired as those of social insects. Human cooperation is frequently based on individual decisions guided by personal interests.

Our communities are no super-organisms. Former Prime Minister Margaret Thatcher pithily claimed that "there is no such thing as society." This can serve as the rallying cry of *methodological individualism*—a research program aiming to explain collective phenomena bottom-up, by the interactions of the individuals involved. The mathematical tool for this program is game theory. All "players" have their own aims. The resulting outcome can be vastly different from any of these aims, of course.

1.2 THE INVISIBLE HAND

If the end result depends on the decisions of several, possibly many individuals having distinct, possibly opposite interests, then all seems set to produce a cacophony of conflicts. In his *Leviathan* from 1651, Hobbes claimed that selfish urgings lead to "such a war as is every man against every man." In the absence of a central authority suppressing these conflicts, human life is "solitary, poore, nasty, brutish, and short." His French contemporary Pascal held an equally pessimistic view: "We are born unfair; for everyone inclines towards himself. . . . The tendency towards oneself is the origin of every disorder in war, polity, economy etc." Selfishness was depicted as the root of all evil.

But one century later, Adam Smith offered another view. An invisible hand harmonizes the selfish efforts of individuals: by striving to maximize their own revenue, they maximize the total good. The selfish person works inadvertently for the public benefit. "By pursuing his own interest he frequently promotes that of the society more effectually than when he really intends to promote it." Greed promotes behavior beneficial to others. "It is not from the benevolence of the butcher, the brewer, or the baker, that we expect our dinner, but from their regard to their own self-interest. We address ourselves, not to their humanity but to their self-love, and never talk to them of our own necessities but of their advantages."

A similar view had been expressed, well before Adam Smith, by Voltaire in his *Lettres philosophiques*: "Assuredly, God could have created beings uniquely interested in the welfare of others. In that case, traders would have been to India by charity, and the mason would saw stones to please his neighbor. But God designed things otherwise. . . . It is through our mutual needs that we are useful to the human species; this is the grounding of every trade; it is the eternal link between men."

Adam Smith (who knew Voltaire well) was not blind to the fact that the invisible hand is not always at work. He merely claimed that it *frequently* promotes the interest of the society, not that it always does. Today, we know that there are many situations—so-called social dilemmas—where the invisible hand fails to turn self-interest to everyone's advantage.

1.3 THE PRISONER'S DILEMMA

Suppose that two individuals are asked, independently, whether they wish to give a donation to the other or not. The donor would have to pay 5 dollars for the beneficiary to receive 15 dollars. It is clear that if both players cooperate by giving a donation to their partner, they win 10 dollars each. But it is equally clear that for each of the two players, the most profitable strategy is to donate nothing, i.e., to defect. No matter whether your co-player cooperates or defects, it is not in your interest to part with 5 dollars. If the co-player cooperates, you have the choice between obtaining, as payoff, either 15 dollars, or 10. Clearly, you should defect. And if the co-player defects, you have the choice between getting nothing, or losing 5 dollars. Again, you should defect. To describe the Donation game in a nutshell:

		if the co-player makes a donation	if the co-player makes no donation
My payoff	if I make a donation	10 dollars	−5 dollars
	if I make no donation	15 dollars	0 dollars

But the other player is in the same situation. Hence, by pursuing their selfish interests, the two players will defect, producing an outcome that is bad for both. Where is the invisible hand? "It is often invisible because it is not here," according to economist Joseph Stiglitz.

This strange game is an example of a *Prisoner's Dilemma*. This is an interaction between two players, player I and II, each having two options: to cooperate (play C) or to defect (play D). If both cooperate, each obtains a *Reward* R that is higher than the *Punishment* P, which they obtain if both defect. But if one player defects and the other cooperates, then the defector obtains a payoff T (the *Temptation*) that is even higher than the Reward, and the cooperator is left with a payoff S (the *Sucker's payoff*), which is lowest of all. Thus,

$$T > R > P > S. \tag{1.1}$$

As before, it is best to play D, no matter what the co-player is doing.

		if player II plays C	if player II plays D
Payoff for player I	if player I plays C	R	S
	if player I plays D	T	P

If both players aim at maximizing their own payoff, they end up with a suboptimal outcome. This outcome is a trap: indeed, no player has an incentive to switch unilaterally from D to C. It would be good, of course, if both *jointly* adopted C. But as soon as you know that the other player will play C, you are faced with the Temptation to improve your lot still more by playing D. We are back at the beginning. The only consistent solution is to defect, which leads to an economic stalemate.

The term "Prisoner's Dilemma" is used for this type of interaction because when it was first formulated, back in the early fifties of last century, it was presented as the story of two prisoners accused of a joint crime. In order to get confessions, the state attorney separates them, and proposes a deal to each: they can go free (as state's witness) if they rat on their accomplice. The accomplice would then have to face ten years in jail. But it is understood that the two prisoners cannot *both* become state's witnesses: if both confess, both will serve seven years. If both keep mum, the attorney will keep them in jail for one year, pending trial. This is the original Prisoner's Dilemma. The Temptation is to turn state's witness, the Reward consists in being released after the trial, (which may take place only one year from now), the Punishment is the seven years in jail and the Sucker's payoff amounts to ten years of confinement.

The young mathematicians who first investigated this game were employees of the Rand Corporation, which was a major think tank during the Cold War. They may have been inspired by the dilemma facing the two superpowers. Both the Soviet Union and the United States would have been better off with joint nuclear disarmament. But the temptation was to keep a few atomic bombs and wait for the others to destroy their nuclear arsenal. The outcome was a horrendously expensive arms race.

1.4 THE SNOWDRIFT GAME

The Prisoner's Dilemma is not the only social dilemma displaying the pitfalls of selfishness. Another is the so-called *Snowdrift* game. Imagine that the experimenter promises to give the two players 40 dollars each, on receiving from them a "fee" of 30 dollars. The two players have to decide separately whether they want to come up with the fee, knowing that if they both do, they can share the cost. This seems to be the obvious solution: they would then invest 15 dollars each, receive 40 in return, and thus earn 25 dollars. But suppose that one player absolutely refuses to pay. In that case, the other player is well advised to come up with 30 dollars, because this still leads to a gain of 10 dollars in the end. The decision is hard to swallow, however, because the player who invests nothing receives 40 dollars. If both players are unwilling to pay the fee, both receive nothing. This can be described

		if my co-player contributes	if my co-player refuses to contribute
My payoff	if I contribute	25	10
	if I refuse to contribute	40	0

as a game with the two options C (meaning to be willing to come up with the fee) and D (not to be willing to do so). If we denote the payoff values with R,S,T, and P, as before, we see that in the place of (equation 1.1.) we now have

$$T > R > S > P. \tag{1.2}$$

Due to the small difference in the rank-ordering (only S and P have changed place), playing D is not *always* the best move, irrespective of the co-player's decision. If the co-player opts for D, it is better to play C. In fact, for both players, the best move is to do the opposite of what the co-player decides. But in addition, both know that they will be better off by being the one who plays D. This leads to a contest. If both insist on their best option, both end up with the worst outcome. One of them has to yield. This far the two players agree, but that is where the agreement ends.

The name *Snowdrift* game refers to the situation of two drivers caught with their cars in a snow drift. If they want to get home, they have to clear a path. The fairest solution would be for both of them to start shoveling (we assume that both have a shovel in their trunk). But suppose that one of them stubbornly refuses to dig. The other driver could do the same, but this would mean sitting through a cold night. It is better to shovel a path clear, even if the shirker can profit from it without lifting a finger.

1.5 THE REPEATED PRISONER'S DILEMMA

The prisoners, the superpowers, or the test persons from the economic experiments may seem remote from everyday life, but during the course of a day, most of us will experience several similar situations in small-scale economic interactions. Even in the days before markets and money, humans were engaged in ceaseless give and take within their family, their group or their neighborhood, and faced with the temptation to give less and take more.

The artificial aspect of the Donation game is not due to its payoff structure, but to the underlying assumption that the two players interact just once, and then go their separate ways. Most of our interactions are with household members, colleagues, and other people we are seeing again and again.

The games studied so far were *one-shot* games. Let us assume now that the same two players repeat the same game for several rounds. It seems obvious that a player who yields to the temptation of exploiting the co-player must expect retaliation. Your move in one round is likely to affect your co-player's behavior in the following rounds.

Thus let us assume that the players are engaged in a Donation game repeated for six rounds. Will this improve the odds for cooperation? Not really, according to an argument called *backward induction*. Indeed, consider the sixth and last round of the new game. Since there are no follow-up rounds, and since what's past is past, this round can be viewed in isolation. It thus reduces to a one-shot Donation game, for which selfish interests, as we have seen, prescribe mutual defection. This is the so-called "last-round effect." Both players are likely to understand that nothing they do can alter this outcome. Hence, they may just as well take it for granted, omit it from further consideration, and just deal with the five rounds preceding the last one. But for the fifth round, the same argument as before prescribes the same move, leading to mutual defection; and so on. Hence backward induction shows that the players should never cooperate. The players are faced with a money pump that can deliver 10 dollars in each round, and yet their selfish interests prescribe them not to use it. This is bizarre. It seems clearly smarter to play C in the first round, and signal to the co-player that you do not buy the relentless logic of backward induction.

It is actually a side-issue. Indeed, people engaged in ongoing everyday interactions do rarely know beforehand which is the last round. Usually, there is a possibility for a further interaction—the *shadow of the future*. Suppose for instance that players are told that the experimenter, after each round, throws dice. If it is six, the game is stopped. If not, there is a further round of the Donation game, to be followed again by a toss of the dice, etc. The duration of the game, then, is random. It could be over after the next round, or it could go on for another twenty rounds. On average, the game lasts for six rounds. But it is never possible to exploit the co-player without fearing retaliation.

In contrast to the one-shot Prisoner's Dilemma, there now exists no strategy that is best against all comers.If your co-player uses an unconditional strategy and always defects, or always cooperates, come what may, then it is obviously best to always defect. However, against a touchy adversary who plays C as long as you do, but turns to relentlessly playing D after having experienced the first defection, it is better to play C in every round. Indeed, if you play D, you exploit such a player and gain an extra 5 dollars; but you lose all prospects of future rewards, and will never obtain a positive payoff in a further round. Since you can expect that the game lasts for 5 more rounds, on average, you give up 50 dollars.

What about the repeated Snowdrift game? It is easy to see that if the two players both play C in each round, or if they alternate in paying the fee, i.e., being the C player, then they will both do equally well, on average; but is it likely that they will reach such a symmetric solution? Should we rather expect that one of the two players gives in, after a few rounds, and accepts grudgingly the role of the exploited C player? The joint income, in that case, is as good as if they both always cooperate, but the distribution of the income is highly skewed.

1.6 TOURNAMENTS

Which strategy should you chose for the repeated Prisoner's Dilemma, knowing that none is best? Some thirty years ago, political scientist Robert Axelrod held a computer tournament to find out. People could submit strategies. These were then matched against each other, in a round-robin tournament: each one engaged each other in an iterated Prisoner's Dilemma game lasting for 200 rounds (the duration was not known in advance to the participants, so as to offer no scope for backward induction). Some of the strategies were truly sophisticated, testing out the responses of the co-players and attempting to exploit their weaknesses. But the clear winner was the simplest of all strategies submitted, namely *Tit for Tat* (*TFT*), the epitome of all retaliatory strategies. A player using *TFT* plays C in the first move, and from then on simply repeats the move used by the co-player in the previous round.

The triumph of *TFT* came as a surprise to many. It seemed almost paradoxical, since *TFT* players can *never* do better than their co-players in a repeated Prisoner's Dilemma game. Indeed, during the sequence of rounds, a *TFT* player is never ahead. As long as both players cooperate, they do equally well. A co-player using D draws ahead, gaining T versus the *TFT* player's payoff S. But in the following rounds, the *TFT* player loses no more ground. As long as the co-player keeps playing D, both players earn the same amount, namely P. If the co-player switches back to C, the *TFT* player draws level again, but resumes cooperation forthwith. At any stage of the game, *TFT* players have either accumulated the same payoff as their adversary, or are lagging behind by the payoff difference $T - S$. But in Axelrod's tournament, the payoffs against all co-players had to be added to yield the total score; and thus *TFT* ended ahead of the rest, by doing better than every co-player *against the other entrants*.

Axelrod found that among the 16 entrants for the tournaments, eight were *nice* in the sense that they never defected first. And these eight took the first eight places in the tournament. Nice guys finish first! In fact, Axelrod found that another strategy even "nicer" than *TFT* would have won the tournament, had it been entered. This was *TFTT* (*Tit for Two Tats*), a strategy prescribing to defect only after two consecutive D's of the co-player. When Axelrod repeated his tournaments, 64 entrants showed up, and one of them duly submitted *TFTT*. But this strategy, which would have won the first tournament, only reached rank 21. Amazingly, the winner of the second tournament was again the simplistic *TFT*. It was not just nice, it was transparent, provokable, forgiving, and robust.This bouquet of qualities seemed the key to success.

1.7 ARTIFICIAL SOCIETIES

The success of Axelrod's tournaments launched a flurry of computer simulations. Individual-based modeling of artificial societies greatly expanded the scope of game theory. Artificial societies consist of fictitious individuals, each equipped with a strategy specified by a program. These individuals meet randomly, engage in an iterated Prisoner's Dilemma game, and then move on to meet others. At the end, the accumulated payoffs are compared. Often, such a tournament is used to update the artificial population. This means that individuals produce "offspring", i.e., other fictitious individuals inheriting their strategy. Those with higher payoffs produce more individuals, so that successful strategies spread. Alternatively, instead of inheriting strategies, the new individuals can adapt by copying strategies, preferentially from individuals who did better. In such individual-based simulations, the frequencies of the strategies change with time. One can also occasionally introduce small minorities using new strategies, and see whether these spread or not. In chapter 2, we shall describe the mathematical background to analyze such models.

Let us consider, for instance, a population using only two strategies, *TFT* and *AllD*. The average payoff for a *TFT* player against another is 60 dollars (corresponding to 6 rounds of mutual cooperation). If a *TFT* player meets an *AllD* player, the latter obtains 15 dollars (by exploiting the co-player in the first round) and the former loses 5 dollars. If two *AllD* players meet each other, they get nothing.

		if the co-player plays *Tit for Tat*	if the co-player always defects
My payoff	if I play *Tit for Tat* (*TFT*)	60	−5
	if I always defect (*AllD*)	15	0

Players having to choose among these two strategies fare best by doing what the co-player does, i.e., playing *TFT* against a *TFT* player and *AllD* against an *AllD* player. But in individual-based modeling, the fictitious players have no options. They are stuck with their strategy, and do not know their co-player's strategy in advance. Obviously, the expected payoff depends on the composition of the artificial population. If most play *TFT*, then *TFT* is favored; but in a world of defectors, *AllD* does better. In the latter case, the players are caught in a social trap. Games with this structure are also known as *Staghunt* games. A fictitious population will evolve towards a state where all play the same strategy. The outcome depends on the initial condition. It is easy to see that if there are more than ten percent *TFT* players around, they will succeed. If the probability of another round is close to 1, i.e., if the expected number of future rounds is large, then even a small percentage of reciprocators suffices to overcome the defectors.

The computer simulations show, however, that a *TFT* regime is not the "end of history." Indeed, *AllC* players can invade, since in a *TFT* world, they do as well as the retaliators. If a small minority of *AllC* players is introduced into a population where all residents play *TFT*, they will do just as well as the resident majority. In fact, under plausible conditions they even offer an advantage. Indeed, an unconditional strategy seems easier to implement than a conditional strategy. More importantly, if a mistake occurs in an interaction between two *TFT* players, either because a move is mis-implemented or because it is misunderstood by the co-player, then the *TFT* players are caught in a costly sequence of alternating defections, in the relentless logic of "an eye for an eye." In computer simulations, such mistakes can be excluded, but in real-life interactions, they cannot. Mis-implementing a move or misunderstanding the co-player's action is common. An *AllC* player is much less vulnerable to errors: a mistake against a *TFT* player, or against another *AllC* player, is overcome in the very next round.

If individual-based simulations are life-like enough to allow for occasional errors, then a *TFT* regime is unlikely to last for long; less stern strategies such as *AllC* can spread. But once a substantial amount of *AllC* players is around, then *AllD* players can take over. The evolutionary chronicles of artificial populations involved in repeated interactions of the Prisoner's Dilemma type are fascinating to watch. The outcome depends in often surprising ways on the range of strategies tested during the long bouts of trial and error provided by the individual-based simulations. One frequent winner is *Pavlov*, a strategy that begins with a cooperative move and then cooperates if and only if, in the previous move, the co-player choose the same move as oneself. In chapter 3, we shall analyze some of the game theoretical aspects behind individual-based simulations.

1.8 THE CHAMPIONS OF RECIPROCITY

The computer tournaments led to a wave of research on reciprocity. But how much of it relates to the real world, as opposed to thought experiments? If *Tit for Tat* is so good, it should be widespread among fish and fowl. Evolutionary biologists and students of animal behavior uncovered a handful of candidates, but no example was universally accepted. It is difficult, in the wild, to make sure that the payoff values (which, in the animal kingdom, are expressed in the currency of reproductive success) do really obey the ordering given by equation (1.1). It is even more difficult to infer, from observing the outcome of a few rounds, which strategy was actually used. *TFT* is but one of countless possibilities. Moreover, many real-life collaborations offer plenty of scope for other explanations, for instance via kin-selection.

Today, after a few decades of this research, the net result is sobering. Beyond the realm of primates, there are few undisputed examples of *Tit for Tat*–like behavior. On the other hand, an overwhelming body of evidence proclaims that humans are, far and wide, the champions of reciprocity. This is not only clear from a huge amount of psychological tests and economic experiments. Brain imaging seems to support the view that part of our cortex is specialized to deal with the ceaseless computations required to keep count of what we give and what we receive, and to respond emotionally to perceived imbalance. Moreover, humans have an extraordinary talent for empathy—the ability to put oneself into another's shoes. Not only do we have an instinctive propensity to imitate another person's acts, we also are able to understand the intentions behind them.

For human nature, retaliation comes easy. The impulse is so strong that little children kick back at inanimate objects that hurt them. More importantly, we empathize with strangers interacting with each other, even as mere bystanders, as so-called *third parties*. This opens up the field of indirect reciprocation.

1.9 ENTER THE THIRD PARTY

You may know the old story about the aged professor who conscientiously attends the funerals of his colleagues, reasoning that "if I don't come to theirs, they won't come to mine." Clearly, the instinct of reciprocation is misfiring here. On second thought, it seems likely that the funeral of the professor, when it comes, will indeed be well-attended. His acts of paying respect will be returned, not by the recipients, but by third parties. This is indirect reciprocity.

In direct reciprocity, an act of helping is returned by the recipient. "I'll scratch your back because you scratched mine." But in indirect reciprocity, an act of helping is returned, not by the recipient, but by a third party. "I'll scratch your back because you scratched somebody else's." This seems much harder to understand. Nevertheless the principle suffices, so it seems, to keep cooperation going—or more precisely, to keep it from being exploited, and thereby ruined.

Indeed, an exploiter will gladly accept help without ever giving anything in return. If all do this, cooperation has vanished. Therefore, such exploitation should be repressed. The obvious way to do this is to withhold help from those who are known to withhold help. This channels cooperation towards the cooperators. But a moment's reflection shows that the principle is not consistent: if you restrain from helping an exploiter, you may be perceived by third parties as an exploiter yourself, and suffer accordingly. But we shall see in chapter 4 that indirect reciprocation can nonetheless hold its own.

If third parties can distinguish between a justified refusal to help an exploiter, and an unjustified refusal, then those who refuse to help exploiters run no risk of being seen as exploiters themselves. Bystanders must be able to assess actions as justified or not, i.e., as good or bad, even when they are not directed at themselves.

A closer investigation reveals that there are many possible assessment norms. Some work better than others. All require a considerable amount of information about the other members of the population. The faculty to process such information may have evolved in the context of direct reciprocity already. It is certainly helpful, before you launch into a series of iterated games, to know how your prospective partners behaved towards their previous co-players. In this sense, indirect reciprocity "may have emerged from direct reciprocity in the presence of interested partners," in the words of evolutionary biologist Richard Alexander. But whereas direct reciprocity requires repetition, indirect reciprocity requires reputation. In the former case, you must be able to recognize your co-players; in the latter, you must know about them. "For direct reciprocity, you need a face; for indirect reciprocity, you need a name" (David Haig).

Subscribers to eBay auctions are asked to state, after each transaction, whether they were satisfied with their partner or not. The ratings of eBay members, accumulated over twelve months, are public knowledge. This very crude form of assessment seems to suffice for the purpose of reputation-building, and seems to be reasonable proof against manipulation. Other instances of public score-keeping abound in social history: a cut thumb signified a thief, a shaved head told of a fallen woman, a medal announced a hero. Reputation mechanisms have also played an important role in the emergence of long-distance trade.

If the community is small enough, direct experience and observation are likely to be sufficient to sustain indirect reciprocity. In larger communities, individuals often have to rely on third-party knowledge. Gossip must always have been the major tool for its dissemination. It may well be that our language instinct and our moral sense co-evolved.

1.10 MORAL SENTIMENTS AND MORAL HAZARDS

The role of moral judgments in everyday economic decisions was well understood by Adam Smith, who wrote his book on *The Theory of Moral Sentiments* even before turning to *The Wealth of Nations*. Later generations of economists tended to neglect the issue of moral emotions. But today, it is generally recognized that our "propensity to trade, barter, and truck" requires, first and foremost, trust. Trust has been hailed as a "lubricant of social life." Different communities operate on different levels of mutual trust. A firm moral basis for economic interactions and a consensual "rule of law" appear to be major indicators for the wealth of nations, more important than population size or mineral resources.

The human propensity to trust is well captured in the so-called Trust game. This is built upon the Donation game: in the first stage, the Donor (or Investor) receives a certain endowment by the experimenter, and can decide whether or not to send a part of that sum to the Recipient (or Trustee), knowing that the amount will be tripled upon arrival: each euro spent by the Investor yields three euros on the Trustee's account. In the second stage, the Trustee can return some of it to the Investor's account, on a one-to-one basis: it costs one euro to the Trustee to increase the Investor's account by one euro. This ends the game. Players know that they will not meet again. Clearly, a selfish Trustee ought to return nothing to the Investor. A selfish Investor ought therefore to send nothing to the Trustee. Nevertheless, in real experiments, transfers are frequent, and often lead to a beneficial outcome for both players. The

Trust game is analyzed in chapter 5, where it is shown that, unsurprisingly, concerns for reputation play a vital role.

Many real-life economic interactions contain elements of the Trust game. For instance, if I entrust money to a fund manager, I expect a positive return; and the fund manager also expects a benefit. The most important asset of a fund is its good reputation. A banker who fails to return the money will meet double trouble. On the one hand, the persons who entrusted him with their money will insist on getting it back; on the other hand, no new clients will be willing to trust him with their earnings. Both direct and indirect reciprocity are at work.

Economists and social scientists are increasingly interested in indirect reciprocity because one-shot interactions between far-off partners become more and more frequent in today's global market. They tend to replace the traditional long-lasting associations and long-term interactions between relatives, neighbors, or members of the same village. A substantial part of our life is spent in the company of strangers, and many transactions are no longer face-to-face. The growth of e-auctions and other forms of e-commerce is based, to a considerable degree, on reputation and trust. The possibility to exploit such trust raises what economists call moral hazards. How effective is reputation, especially if information is only partial?

Evolutionary biologists, on the other hand, are interested in the emergence of human communities. A considerable part of human cooperation is based on moralistic emotions, such as, for instance, anger directed towards cheaters, or the proverbial "warm inner glow" felt after performing an altruistic action. It is intriguing that humans not only feel strongly about interactions that involve them directly, but also about actions between third parties. They do so according to moral norms. These norms are obviously to a large extent culture-specific; but the *capacity* for moral norms appears to be a human universal for which there is little evidence in other species.

It is easy to conceive that an organism experiences as "good" or "bad" anything that affects its own reproductive fitness in a positive or negative sense. Our pleasure in eating calorie-rich food or experiencing sex has evolved because it heightens our chances of survival and reproduction. In the converse direction, disgust, hunger, and pain serve as alarm signals helping us to avoid life-threatening situations. The step from there to assessing actions between third parties as "good" or "bad" is not at all obvious. The same terms "good" and "bad" that are applied to pleasure and discomfort are also used in judging interactions between third parties: this linguistic quirk reveals an astonishing degree of empathy, and reflects highly developed faculties for cognition and abstraction.

1.11 ULTIMATUM EXPERIMENTS

A series of economic experiments documents that indirect reciprocity works. The more the players know about each other, the more they are likely to provide help to each other. There seems clear evidence for the player's concern with their own reputation. But interestingly, many players also tend to help, although to a lesser degree, when they know that nobody can watch them and that their action will not affect their reputation. Moreover, they are more likely to give help if they have previously received help. This is difficult to explain through self-interest. It could be the outcome of a maladaptation. If somebody holds a door open for you, then you are more likely to hold the door open for the next person, motivated by a vague feeling of gratitude. It may well be that similar reflexes of misdirected reciprocity operate in other social and economic contexts.

A particularly revealing light on our propensity to empathize with others is provided by the Ultimatum game. In this experiment, two anonymous players are randomly alloted the role of Proposer and Responder. The Proposer is then given 10 euros, and asked to divide that amount between the two players, subject to the Responder's acceptance. Thus if the Responder accepts the proposed split, then the money will be shared accordingly, and the game is over. But if the Responder rejects the offer, then the game is also over; the experimenter withdraws the 10 euros, and both players receive nothing. This is it: no haggling, and no second round.

It seems obvious that the Responder should accept any positive offer, since this is better than nothing. Accordingly, a selfish Proposer should offer only a minimal share. In real experiments, however, most players offer a fair split—something between forty and fifty percent of the total. On the few occasions that less than twenty percent is offered, the Responder usually refuses. Proposers seem to anticipate this.

In most cases, refusals are correlated with angry feelings. Brain imaging shows that unfair offers elicit activity in two brain areas: one is in the left frontal part of the brain, which is usually associated with rational decisions, while the other is much deeper, in the striatum, which is linked with emotional responses. The tug of war between these two parts of the brain corresponds to the tension between (a) accepting the low offer, on the grounds that it is better than nothing, and (b) telling the unfair Proposer to go to hell. The intensity of the brain activities in the two locations foretells the decision, even before the Responder is aware of it.

The Ultimatum game experiment has been repeated many times. A large number of variants have been explored. For instance, if the Proposer is a computer, the Responder feels no qualms in accepting a small offer. If a game of skill (rather than the toss of a coin) decides who of the two players is going to be the Proposer, then smaller offers are more likely to be accepted: it is as if the Proposer had earned the right to keep a larger part of the sum. Furthermore, if several Responders compete, the Proposer knows that a small offer has a good chance of being accepted.

1.12 FAIRNESS NORMS

An extensive research program has used the Ultimatum game to study fairness norms in many small scale societies, including hunter-gatherers, nomads, slash-and-burn farmers, etc. The average offer varies between cultures. Remarkably, offers in large cities are among the fairest; Mother Nature's son is not always as noble as a city slicker or even an economics undergraduate. But the average offer is always far from the theoretical minimum. Norms of fairness seem wide-spread, maybe universal. How did they emerge?

Again, one possible explanation relies on reputation. Once it becomes known that you reject unfair offers, people will think twice before proposing them to you. The long term benefit of rejecting the offer may well outweigh the loss, which is all the smaller, the smaller the share you have been offered. In chapter 5, a simple mathematical model reveals how concerns for reputation can lead to the establishment of fairness norms. Paradoxically, this works only if Proposers who, ordinarily, are willing to offer a fair share, do occasionally yield to the temptation of offering less if they can get away with it. It is thus precisely when fairness norms are not hard-wired, and may be overcome by the opportunistic urgings of selfishness, that these norms are upheld in the population.

What have real experiments (as opposed to individual-based computer simulations) to say about this? It is easy to set up two distinct treatments of the Ultimatum game, each with a large population of anonymous test subjects who are randomly paired. In one treatment, players play the game for ten rounds (always against different co-players, of course) and nobody knows anything about the outcome of the previous rounds. In the other treatment, the outcomes are known to all. It is obviously only in the second treatment that players can hope to establish a reputation for rejecting small offers. The outcome is clear: the unfair offers tend to be considerably rarer. It is as if the Proposers anticipate that Responders fear to get exploited if it becomes known that they have meekly consented to a trifling share.

If Responders, in the Ultimatum game, reject an unfair offer, they have every interest in letting this be known to others. Under natural circumstances, an emotional response is likely to attract some attention. Anger is loud.

This being said, the fact remains that Ultimatum offers are often fair even if players know that the outcome will be kept secret. This seems puzzling. But it could well be that the players' subconscious is hard to convince that nobody will ever know. In our evolutionary past, it must have been exceedingly difficult to keep secrets from the small, lifelong community of tribal members and village dwellers in which our ancestors lived. Moreover, the belief of an overwhelming majority in a personal god watching them day and night shows that the feeling of being observed is deep-rooted and wide-spread.

Psychologists have devised ingenious experiments to document that our concern of being observed is easily aroused. For instance, players sitting in a cubicle in front of a computer are strongly affected by the mere image of an eye on the computer screen. They know that the eye is purely symbolic, but nevertheless they react to it. In another wonderfully simple experiment, the mere picture of eyes on a cafeteria wall next to the "honesty box" in a British university department sufficed to raise the amount staff members paid for coffee and cookies by more than two hundred percent. Obviously, it is easy to trigger a concern about being watched. And it is worth emphasizing that in our species, the eyes are uniquely revealing: due to the white color around the iris, the direction of their gaze is particularly noticeable. Incidentally it seems that test persons react the same, whether one or several persons are watching. This shows that they believe, at least subconsciously, that news will spread through gossip. One witness is enough.

1.13 PUBLIC GOODS GAMES

The games considered so far, such as Prisoner's Dilemma, Snowdrift, Trust, or Ultimatum, are two-person games. But many economic interactions involve larger groups of actors. The notion of reciprocation becomes problematic, in such cases. If your group includes both cooperators and defectors, whom do you reciprocate with? This introduces a new twist to social dilemmas.

So-called Public Goods games offer experimental instances of such dilemmas. Here is a typical specimen of such an experiment: Six anonymous players are given 10 dollars each, and are offered the opportunity to invest some of it in a common pool. The players know that the content of the common pool will subsequently be tripled by the experimenter, and that this "public good" will then be divided equally among all six players—irrespective of the amount that they contributed.

Obviously, all players are well off if they fully invest their 10 dollars. They receive 30 dollars each. But if one player invests nothing, and the others contribute fully, then each of the six players receives 25 dollars back from the public good; the defector, who contributed nothing, and thus kept the initial 10 dollars, ends up with a net sum of 35 dollars, 10 dollars more than the others.

For each dollar invested, only 50 cents return to the contributor. A selfish income-maximizer ought to invest nothing. But if all players do this, they have missed a first-class opportunity to increase their stocks.

In real experiments, most players invest on average half their initial amount, or even more. There are considerable variations among the individual contributions, but many players seem to hedge their bets. However, if the game is repeated for a few rounds, the contributions decline from round to round, and may end up at zero. The mechanism seems clear. If players notice that they have contributed more than others, they feel exploited, and reduce their future investments. But this causes other cooperators to feel exploited, and they reduce their contribution in turn. Cooperation goes down the drain.

In the repeated Prisoner's Dilemma game, a strategy like *Tit for Tat* allows one to retaliate against defectors. Such a reciprocating strategy loses its clout in a repeated Public Goods game. Indeed, by withholding your contribution, you hit friend and foe alike: your response is not directed against defectors only, but affects all the participants of the Public Goods game.

In economic life, similar interactions based on joint efforts, or joint investments, abound. This social dilemma is often described as multi-person Prisoner's Dilemma, or Free-Rider problem, or Tragedy of the Commons. A commons is a piece of grazing land that can be used by all villagers. The tragedy of the commons is due to the fact that it is usually over-exploited, and therefore ruined through overgrazing. Today, there are not many commons left, but the tragedy is still with us: the oceans are our new commons. On a smaller scale, the tragedy can be seen in most communal kitchens. Joint enterprises and common resources offer alluring prospects for cheaters and defectors.

1.14 PUNISH OR PERISH?

If you try riding free on public transportation, or dodging taxes, or littering parks, you run the risk of being caught and fined. Many judicial and legal institutions, as well as moral pressure, aim at keeping our contributions up. Thus the free-rider problem has an obvious solution: cooperation can be bolstered through incentives, by punishing or rewarding individual players, conditional on their behavior. However, legal institutions require a fairly advanced society.

It turns out that in the absence of such institutions, individuals are often willing to make the job of sanctioning their own.This has been neatly demonstrated by a series of experiments. After a round of the Public Goods game, players are told what their co-players contributed, and are given the opportunity of punishing them. If players are punished, this means that a fine of three dollars is deducted from their account. This fine is collected by the experimenter, and does not end up in the pockets of the punishing player. On the contrary, the punishing player must pay one dollar to inflict the punishment: this can be viewed as a fee which has to be paid to the experimenter. The fee is meant to reflect the fact that punishing another individual is usually a costly enterprise: it demands time and energy, and often enough entails some risk.

In the economic experiments, players are often willing to punish, despite the cost to themselves. This seems to be anticipated by most participants. The average level of contributions is higher, with the threat of punishment, than without. Most significantly, if the game is repeated for several rounds (each consisting of a Public Goods game followed by an opportunity for meting out punishment), then the contributions increase from round to round, up to remarkably high levels, see figure 1.1. Punishment obviously boosts the level of cooperation.

Why do people engage in costly punishment? The first explanation is obvious. By punishing defectors, one can hope to reform them. Thus punishers can expect to recoup their fee in the following rounds, through the heightened contributions of the castigated players. But this appears to be only part of the answer. Indeed, in a variant of the game requiring a large population of test persons, the Public Goods groups of six players are newly formed between rounds, and players know that they will never meet a co-player twice. By inflicting punishment, they can possibly turn a defector into a cooperator. However, punishers know that the future contributions of such an improved player will exclusively benefit others. Punishment appears as an altruistic act.

This is a stunning outcome. Without sanctions, the public good, i.e., the tripling of the endowment, is not realized. With sanctions, it is, although selfish reckoning prescribes that costly punishment should not be delivered. In the absence of institutions, some players are willing "to take the law into their own hands," which is also known as "peer-punishing."

What motivates players to punish defectors? Probably, we need not invoke any drive beyond the prevailing tendency to reciprocate. If the players themselves feel exploited, direct reciprocation is at work. If others are exploited, it is indirect reciprocation: humans are often willing to retaliate on behalf of third parties.

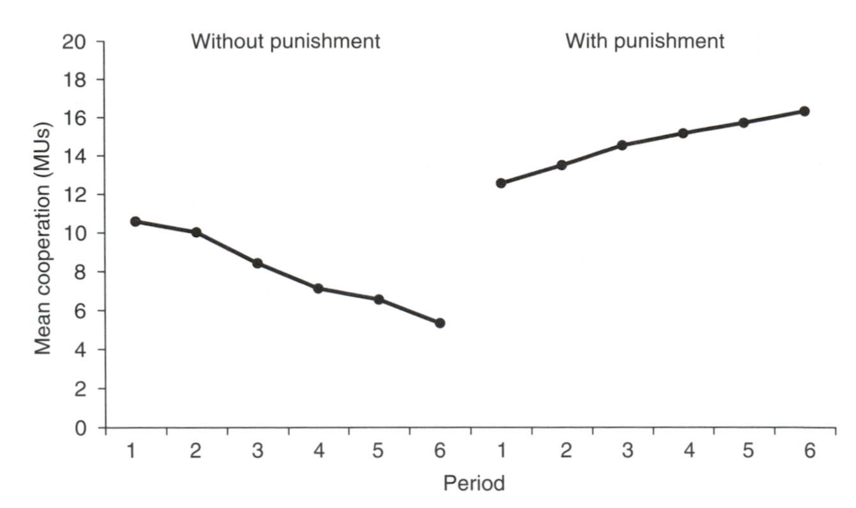

Figure 1.1 Public Goods and punishment. In each of the twelve rounds of the Public Goods game, groups of four players are formed (out of a population of 240 players). The players receive 20 monetary units (MU) per round, and have to decide how much of it to invest, knowing that their contributions will be multiplied by two and divided equally among the four participants. In the treatment "with punishment," players can fine their co-players; fines are collected by the experimenter. Imposing a fine of three monetary units costs a fee of one monetary unit. Players know that they encounter a co-player only once. Shown is the average contribution to the public good in each round. (After Fehr and Gächter, 2002.)

1.15 SECOND-ORDER FREE-RIDING

It is clear that in a population consisting of players ready to punish exploiters, defection makes no sense. The gain from not contributing is more than off-set by the expected fines. Defectors would have to bear the full brunt of punishment from the majority. If punishers (i.e., players who contribute, and impose fines on those who do not contribute) are established in a population, they can resist defectors and uphold cooperation.

But a population of punishers can be subverted by players who contribute, but do not punish. Newcomers of that type do just as well as the resident punishers and thus can slowly spread through random fluctuations. In fact, if occasionally some defectors enter the population, to be promptly assailed by the punishers, then the newcomers would do better than the punishers, by economizing on the cost of punishment. This new type is a second-order exploiter, free-riding on the sanctions delivered by the punishers. Hence, this type will spread: and this means that eventually, there will be too few punishers around to keep the defectors at bay.

Sanctioning can be seen as a service to the community, i.e., a public good. In the long run, second-order exploiters sabotage the enforcement of contributions to the Public Goods game, and therefore both types of contributors—the punishers and the second-order exploiters themselves—will eventually be displaced by defectors.

A remedy coming to mind is "second-order punishment." It consists in punishing not only the "first-order exploiters" who fail to contribute, but also the "second-order exploiters" who contribute, but fail to punish. However, this could in turn give rise to "third-order exploiters" and so on. If punishers of a sufficiently high order dominate the population, there will be few defectors, and therefore few occasions for the lower-order punishers to reveal their limitations to their sterner brethren. Hence, their number can increase through random fluctuations, thus eroding the system.

It seems that reputation, once more, can come to the rescue. Players are less likely to yield to the temptation to cheat if they know that their group includes some punishers. Thus, there is an advantage in being known to react emotionally against exploiters. This will be analyzed in chapter 5. The situation is quite similar to the Ultimatum game. In fact, a Responder who refuses an unfair offer is effectively punishing the Proposer. The more unfair the offer, the less is the cost to the punisher, and the heavier the fine to the punished player.

A similar mechanism operates with positive incentives. If players, after a Public Goods round, are given the possibility to reward the high contributors, at a cost to themselves, they are able to promote the tendency to cooperation. Again, this system is threatened by those players who contribute, and thus benefit from rewards, but do not reward others. Such players are obviously free-riding at the expense of the rewarders, and can subvert the incentive-based system. But if rewarding players can acquire a reputation, they are more likely to experience high levels of cooperation in their group.

The similarity between reward and punishment stops at a point. In a society where everyone contributes, punishers have nothing to do, but rewarders have to dig deep into their pockets. In this sense, punishment is more efficient. Ideally, the mere threat suffices. In real experiments, however, one often finds that while fines do certainly increase cooperation, they may be so expensive that the average payoff in the group is actually lower than in the less cooperative groups playing the Public Goods game without punishment, at least during the first few rounds. Moreover, in many societies asocial punishment (i.e., the punishment of do-gooders) is frequent, and thus throws a spanner in the work of sanctions to uphold cooperation.

1.16 VOLUNTARY PARTICIPATION

Even granted that reputation can stabilize a population of punishers, there remains the problem of explaining how sanctioning can emerge. In a world of defectors, punishers would have to punish right and left. Their payoff would be low and their behavior unlikely to catch on.

This is different, however, if players are not obliged to participate in the Public Goods game, and can opt out of it if they wish. This situation seems natural enough. In many cases, individuals can decide whether to play the game or not. In town, you need not use public transportation: walking is fine. In a hunter-gatherer tribe, you need not join the big-game hunt, or the raiding party, if you suspect that the other participants are laggards. Collecting mushrooms or fruits can provide an option that makes you independent from the others. You need no assistance.

Suppose thus that there exists an alternative to participation in the joint effort, an alternative whose outcome does not depend on what the others are doing. We may then see the Public Goods game as a newly arising opportunity. A mammoth has moved into your valley. Will it pay to join the hunt? Participating in the common effort means effectively to bet on cooperation. We shall assume that if all participants contribute, engaging in the Public Goods game is more profitable than the alternative of not participating; but that if the other participants do not contribute, the Public Goods game is a waste of time that ought to be avoided. Searching for mushrooms is more promising, in that case.

Let us first consider this "optional Public Goods game" without punishment. In that case, the three strategies (to contribute, to defect, or to abstain) are superseding each other in a cyclic fashion, as in the familiar Rock-Paper-Scissors game. If the population consists mostly of cooperators contributing to the joint effort, then it is best, from the selfish viewpoint, to exploit them. But if most players switch to defection, then the Public Goods game is unprofitable and it is better not to participate at all. Finally, if most players are not participating, then cooperation is the best option. This last statement may seem surprising. But if few players are willing to participate, then most teams will be small, and in this case cooperators can do better, on average, than defectors, despite the fact that within each team, defectors do better than cooperators.

These Rock-Paper-Scissors cycles, from contributing to defecting to abstaining to contributing again, do not yet suffice to establish cooperation. In the long-term average, the payoff is not higher than the payoff for non-participants. But as we shall see in chapter 6, if the option of punishing the exploiters is added, then cooperation will be established for most of the time. This is a statistical result. Under stochastic fluctuations, punishers can be subverted after some time by second-order exploiters, and these in turn by defectors; but after such a break-down of cooperation, punishers re-emerge. In the long term, they dominate the population.

The outcome seems paradoxical. In interactions requiring a joint effort, cooperation based on coercion can emerge and prevail, but only if the participation is voluntary. If participation is compulsory, coercion fails and defectors win.

Several economic experiments support the validity of this theoretical conclusion. In Prisoner's Dilemma games and Public Goods games, cooperation is more likely to be achieved if players have the option to abstain from the game. In one particularly telling experiment, players from a large pool had the possibility to choose, between rounds, not whether to participate in a Public Goods game or not, but whether to play their Public Goods game with or without the punishment option. In the first round, most players decided against the version with punishment. This seems understandable. Nobody wants to be punished, and many people dislike punishing; but by the fifth or sixth round, almost all players had switched, on their own free will, to the version with punishment, and cooperated assiduously. They effectively "voted with their feet" for the threat of sanctions, understanding that it made cooperation more likely. This experiment looks almost like a morality play, illustrating the philosophy of the social contract.

1.17 THE GENTLE ART OF ABSTRACTION

How relevant are economic experiments? Often, their most striking aspect is a stark artificiality. They are remote from everyday experience.

This in itself need not be a weakness. Classical experiments in physics or physiology are equally remote from everyday life. Their aim is to probe nature, not to mimic it. It can be argued, for instance, that a major asset of the Ultimatum game consists in creating a situation that players have never encountered before. We have been exposed to haggling, to the rituals of offer and counteroffer, and to market competition. The barren "take it or leave it" alternative of the Ultimatum is profoundly alien to most players. By catching us on the wrong foot, the experimenter forces us to decide spontaneously, rather than rely on force of habit.

The anonymity under which most economic experiments are performed excludes all possible effects of relatedness, reputation, future interactions, or advertising. Anonymity is not a condition that humans have often encountered in their evolutionary past. Most of human evolution took place in small tribes and villages, with everyone knowing everything about everyone else. We have certainly not been adapted, in our evolutionary past, to transferring small sums of money under contrived rules to faceless strangers. It makes no sense to assume that Ultimatum games or Trust games, in their clinical sterility, have shaped our evolution. But human behavior is based on evolved traits, and by varying the treatments in economic experiments, we may hope to unveil these traits.

For instance, players who are allowed to briefly talk with each other, before engaging in a Prisoner's Dilemma game, are more cooperative. Moreover, they can predict very accurately, after a short conversation, whether their co-players will cooperate or not. Even without knowing which type of experiment is in store for them, they quickly pick up the relevant clues for summing up their partner. By varying the nature of their conversation, which can be face-to-face, via monitor, through a phone, or merely a brief visual contact, experimenters can hope to understand how we go about assessing strangers.

To give another example, players tend, as we have seen, to reject unfair offers more readily if they know that this becomes known to their future co-players. Nevertheless, even players who are assured that nobody will know about their decision frequently turn small offers away. It would be naive to overlook the possibility that even if players are convinced that nobody is watching, and have grasped the niceties of double-blind experiments, their subconscious may yet harbor some misgivings. We are far from completely understanding when and why subliminal factors can affect decision making. Players can strongly react to an appropriate cue even when knowing that reality does not back it up. An often mentioned example is the sexual arousal produced by erotic magazines.

Experimental game theorists know this, of course. They do not try to reproduce real life interactions, with their plethora of psychological and cultural effects, but aim to dissect the strategic situation down to the bones. Most economic interactions take place with innumerable side-conditions, among people bound by a plethora of ties of personal history and cultural constraints. Experiments must abstract from all these factors.

1.18 HUMAN BEHAVIOR RESUMED IN TWO SECONDS

In a similar spirit of self-imposed limitation, the mathematical models filling most of this book omit all psychological factors but one: selfishness. This by itself need be no severe restriction: most psychologists would agree that it is a good first approximation. (To quote Jonathan Haidt: "If you are asked to explain human behavior in two seconds or less, you might want to say 'self-interest'."). Some very interesting and plausible theoretical approaches assume that individual utilities include the utilities of other players—that equity, for instance, is deemed desirable—but this eminently psychological issue cannot be tackled here. It seems likely that our preferences emerged through evolution, and that a direct path led from the "selfish gene" to human kindness, but such a topic is way beyond the scope of this book, which merely explores, by mathematical means, how selfishness can overcome social dilemmas.

This is not meant to endorse the idea that our social interactions are governed by some "homo economicus" residing in our breast, who calculates strategies to maximize our own gain with cold rationality. According to John Maynard Keynes, economic decisions are often governed by "animal spirits and spontaneous optimism" and depend on "nerves and hysteria and even digestion and reactions to the weather."

Emotions and instincts act as a system of heuristics to guide us through computations which vastly overtax our rational faculties. Similarly, tennis players manage to compute the trajectory of a ball with a speed and precision that no robot can match. The players work it out subconsciously, and it is doubtful whether any Wimbledon winner would become a better player by a course in physics. In an analogous way, we need no pen and paper to figure out our self-interest in practical life. Game theory may, like a course in physics, provide understanding, but it need not furnish recipes for success.

Just as we concentrate, in the following pages, on self-interest as guiding motivation, we will also purposely ignore the effect of social structure. Neglecting networks may be an even more serious distortion of real life than neglecting altruism. The short last chapter 7 provides a brief glimpse at some factors that are left out in all the preceding chapters: namely family ties, neighborhood effects, and group benefits.

The major part of this book thus deals with simple games of cooperation played by selfish individuals in well-mixed, and usually large, populations. This is an admittedly artificial scenario, but our world seems to evolve towards it. Is it the way of the future? It certainly was no part of our evolutionary past. Nothing prepared us for big city life, but we do have an uncanny talent for mixing with strangers and enjoying "the tumultuous sea of human heads," like the nameless hero of Edgar Allan Poe's short story, "The Man of the Crowd."

1.19 FOOD AND MORALS

Most economic experiments use real money, and some critics say this is about all that is real about them. But in fact, a large amount of everyday economic cooperation involves no money at all. We can lend a hand, or provide some information, or share a meal: in each case the psychological feeling is different. To use money, in experiments, is a simple, clear-cut way to reduce framing effects that complicate the strategic issue.

Nevertheless, it is obvious that this way of standardizing outcomes can sometimes be seriously misleading. For instance, when you had been thinking through the alternatives of the Donation game in section 1.3, you probably felt uneasy about one scenario. If the other player trusts you, would you be willing to defect? Most people balk at that point. It usually feels bad to let another person down. The discomfort seems hardly worth the few extra dollars. Indeed, many actual experiments indicate that a majority of players are willing to cooperate. A comforting amount of people are "good-natured." But where does this good nature come from?

Similar questions are raised by the Ultimatum game. Most Proposers offer close to half of the sum and claim that it just seems the fair thing to do. Conceivably, they are fooling themselves, and are simply afraid that a lower offer may be rejected. But why do Responders reject a small offer? Most claim that they are angered by the obvious injustice of the unfair offer. Again, they possibly mistake their own motivation, and are simply anxious to avoid the reputation of being spineless wimps, a reputation that would harm them in the long run. These selfish imputations seem to fail, however, in a variant of the Ultimatum game, which is known as the Dictator game. In this variant the Proposer makes the offer, and the Responder has no say at all: "Dictators" can do as they like.

In the Dictator game, the offers are usually lower than with the Ultimatum game. Nevertheless, a substantial part of the Proposers offers a positive amount. It seems difficult, in this case, to dismiss "good nature." Proposers simply feel that to be generous makes them happy. If, in another twist, Responders in the Dictator game cannot reject the offer, but can write a short note to let the Responder know what they think of it, then offers jump to almost the same level as with the Ultimatum game. Obviously, people do not like to incur the wrath of others, even if that wrath is guaranteed to be completely ineffective. Meting out the purely symbolic punishment of a censorious message is not very costly in that case.

Are we simply afraid of being cursed? It has been argued that a strong motive for cooperation and moral behavior is the fear of punishment by supernatural spirits. Superstitious maladaptations are widespread, possibly because they strongly promote conformism and obedience—traits which often have some survival value.

If we enjoy sex and food, it is because such emotions promote our survival and reproduction. Similarly, our survival and reproduction depends on being successful cooperators, and this is why we enjoy being virtuous, and why we feel that revenge tastes sweet. Moralistic emotions—the warm inner glow of feeling kind, the anger directed at unfair persons, the guilt and shame after committing a reprehensible deed—are deeply anchored in our nature. Moral rules differ among cultures, but juveniles' ability to pick up prevailing norms and make them their own seems to be as much a part of universal human nature as juveniles' ability to pick up and speak the language of their community.

Not all morality is meant to promote altruism and cooperation. Norms of personal cleanliness and purity have a similar ethical status, without having an economic background. But a large part of moral norms serve ultimately to promote the ceaseless give and take that is such an essential part of human behavior. The German playwright Bertold Brecht wrote in his *Threepenny Opera*: "Food comes first, then morals." It is exactly the reverse. Without morals, we could not subsist.

But fortunately, we need not end with homilies. For what they are worth, the simple models analyzed here also contain some more surprising lessons: for instance, that the instinct of revenge, frowned upon as base, can play a useful economic role by deterring defectors; or that our selfish urge to exploit others whenever we can get away with it, keeps retaliators in the population, thus boosting common welfare; or that the option to abstain from a team effort when it appears unpromising actually helps in enforcing team-wise cooperation. Again and again, we find that traits rendering individuals less than perfect uphold social cohesion. So even if you cannot always satisfy your selfish interests, you may find consolation in the thought that they are furthering the common good.

That human and all-too-human foibles and errors sustain cooperation is not new, by the way. It is known as Mandeville's paradox. The author of the *Fable of the Bees* subtitled his work with the slogan: "Private Vices, Publick Benefits." Private selfishness can promote the public good. The "invisible hand" performs surprising tricks.

1.20 REFERENCES

Basic texts on the evolutionary biology of cooperation can be found in the works of Hamilton (1996) and Trivers (2002), see also Trivers (1985), Frank (1998), and Nowak (2006a). Popular expositions are given by Dawkins (1989), Sigmund (1995), and Ridley (1997). For game theoretical descriptions of social dilemmas with minimal technical fuss, see Colman (1995), Binmore (1994), Sugden (1986), Ostrom (1990), and Skyrms (2004). Good surveys on social dilemmas can also be found in Dawes (1980), Cross and Guyer (1980), Heckathorn (1996), Kollock (1998), and Levin (1999). The Tragedy of the Commons and the dilemmas surrounding collective action were presented by Hardin (1968) and Olson (1965). A popular account of the Prisoner's Dilemma is provided by Poundstone (1992). The Prisoner's Dilemma first mention in a textbook goes back to Luce and Raiffa (1957), see also Schelling (1978).The first full book devoted to the game is by Rapoport and Chammah (1965). Rapoport submitted the Tit for Tat strategy to Axelrod's tournaments (Axelrod 1984). Indirect reciprocity can be traced back to Alexander (1987) and Ellison (1994). It was modelled by Nowak and Sigmund (1998a,b), and early experimental tests are described in Wedekind and Milinski (2000) and Wedekind and Braithwaite (2002). The Snowdrift game is due to Sugden (1986), see also Doebeli, Hauert, and Killingback (2004). The Trust game was first proposed by Berg, Dickhaut, and McCabe (1995), the Ultimatum game by Güth, Schmittberger, and Schwarze (1982).The Ultimatum's use for investigating small-scale societies is covered, e.g., in Henrich (2006). The role of reputation in economics is studied, e.g., in Kreps and Wilson (1982) or Kurzban, DeScioli, and O'Brien (2007). Observer effects were observed in Haley and Fessler (2005), Bateson, Nettle, and Roberts (2006) and Burnham and Hare (2007). The role of sanctions in Public Goods games was studied by Yamagashi (1986) and Fehr and Gächter (2000, 2002), see also Boyd and Richerson (1992), O'Gorman, Wilson, and Miller (2005), Gardner and West (2004) and Sigmund (2007). The troubling aspects of antisocial punishment have been uncovered by Herrmann, Thöni, and Gächter (2008). An experiment by Gürerk, Irlenbusch, and Rockenbach (2006) shows that players of Public Good games opt for the possibility of sanctioning defectors, but only after some experience. Positive and negative incentives to promote cooperation have been compared in many investigations, see e.g., Baumeister et al. (2001), Dickinson (2001), Andreoni, Harbaugh, and Vesterlund (2003), Walker and Halloran (2004), or Sefton, Shupp, and Walker (2007). The role of voluntary participation is studied by Orbell and Dawes (1993), Hauert et al. (2002a, 2002b), and Fowler (2005a). Extensive monographs on experimental economics and behavioral games are by Kagel and Roth (1995) and by Camerer (2003), see also Camerer and Fehr (2006) and, for experiments under more natural conditions, Carpenter, Harrison, and List (2005). Theoretical, sociological and psychological studies on ethical norms and morals are by Yamagishi, Jin, and Kiyonari (1999), Bendor and Swistak (2001), Price, Cosmides, and Tooby (2002), Ostrom and Walker (2003), Cose (2004), Kurzban and Houser (2005), Hauser (2006), and Haidt (2007). Human universals are treated in Brown (1991). Richerson and Boyd (2005) offer a unified treatment of cultural and biological evolution. The role of gossip is highlighted by Dunbar (1996) and Sommerfeld et al. (2007). A seminal text on the economic importance of emotional committment is by Frank (1988). The neural basis of emotions related to economic games is studied in Rilling et al. (2002), Sanfey et al. (2003), Fehr (2004), and de Quervain et al. (2004). The importance of reciprocity is stressed by Charness and Haruvy (2002) and Sachs et al. (2004). Bowles and Gintis (2002), see also Gintis et al. (2003, 2005), present an influential approach termed "strong reciprocity": for a critical view, see Burnham and Johnson (2005). Fehr and Schmidt (1999) and Bolton and Ockenfels (2000) show how to interpret experimental outcomes by modifying utilities, so as to incorporate concerns for equity and fairness. There exists a huge literature on economic and social interactions in non-human primates, see e.g., de Waal (1996), Brosnan and de Waal (2003), Stevens, Cushman, and Hauser (2005), Silk (2006), or Warneken and Tomasello (2006). Various forms of punishment in biological communities are covered in Clutton-Brock and Parker (1995), Kiers et al. (2003), or Wenseleers and Ratnieks (2006); for other ways of repressing competition, see Frank (2003).

GAME ZOO: A BRIEF LEXICON OF TWO-PERSON GAMES

Many experimental two-person games are related to social dilemma issues. Typically, the players are anonymous, and are endowed with a certain amount of money beforehand (e.g., a show-up fee). They are asked to make their decision after having understood the rules of the game and being assigned to the role of Proposer and Responder (or Donor and Recipient).

Donation: in some sense, an atom of social interaction. The Donor decides whether to pay one dollar to give a benefit of three dollars to the Recipient.

Prisoner's Dilemma: the mother of all cooperation games is played in many variations. In one particularly transparent set-up, both players engage in a Donation game with each other. When players decide simultaneously, this is similar to a two-player Public Goods game. If both cooperate by sending a gift to the other, both gain two dollars. But sending a gift costs one dollar, so that the best reply to whatever the co-player decides is to send no gift (i.e., to defect). If both players defect, however, they gain nothing.

Ultimatum: the experimenter assigns a certain sum, and the Proposer can offer a share of it to the Responder. If the Responder (who knows the sum) accepts, the sum is split accordingly between the two players, and the game is over. If the Responder declines, the experimenter withdraws the money. Again, the game is over: but this time, neither of the two players gets anything.

Dictator: same as Ultimatum, except that the Responder cannot reject the offer.

Trust: in a first stage, the Proposer (or Investor) can give a certain benefit to the Responder (or Trustee), as in the Donation game. In the second stage, the Responder can decide how much of it to return to the Proposer. This is similar to the sequential Prisoner's Dilemma game (when first one player acts as Donor and then the other).

Repeated Prisoner's Dilemma: the two players interact for several rounds of the Prisoner's Dilemma. Usually, they are not told beforehand when the interaction will be over, so as to avoid "last round effects" (defection motivated by the fact that the co-player cannot retaliate in a one-shot Prisoner's Dilemma game).

Indirect Reciprocity: in a large population of players, two players are sampled at random and play the Donation game or the (one-shot) Prisoner's Dilemma game. This is repeated again and again. The players know that they interact only once, so that retaliation is impossible.

Snowdrift: two players each receive an endowment, on provision that they pay a fee to the experimenter that is lower than the endowment. They must decide whether they are willing to pay the fee or not, knowing that if both are willing, each of them pays only half.

Chapter Two

Game Dynamics and Social Learning

2.1 GAMES

It can be difficult to decide what is best. The task can be fraught with uncertainties (as when an investor wants to optimize a portfolio), or it can be computationally demanding (as when a traveling salesman has to find the shortest route through 87 towns). A peculiar complication arises in interactions between two (or more) decision-makers with different views about what is best. This is the realm of game theory.

As an example, consider two players I and II engaged in the following, admittedly childish game. At a given signal, each holds up one or two fingers. If the resulting sum is odd, player I wins. If the sum is even, player II wins. Clearly there is no outcome that can satisfy both players. One of the players would always prefer to switch to the other alternative. Situations with a similar structure abound in social interactions.

Let us formalize this. Suppose that player I has to choose between n options, or *strategies*, which we denote by $\mathbf{e}_1, \ldots, \mathbf{e}_n$, and player II between m strategies $\mathbf{f}_1, \ldots, \mathbf{f}_m$. If I chooses $\mathbf{e}_i$ and II chooses $\mathbf{f}_j$, then player I obtains a *payoff* a_{ij} and player II obtains b_{ij}. The game, then, is described by two $n \times m$ payoff matrices A and B: alternatively, we can describe it by one matrix whose element, in the i-th row and j-th column, is the pair (a_{ij}, b_{ij}) of payoff values. The payoff is measured on a utility scale consistent with the players' preferences. In biological games, it can be some measure of Darwinian fitness reflecting reproductive success. For simplicity, we stick to monetary payoffs.

The two players in our example could bet, say, one dollar. Each player has two strategies, *even* and *odd*, which correspond to $\mathbf{e}_1$ and $\mathbf{e}_2$ for player I and $\mathbf{f}_1$ and $\mathbf{f}_2$ for player II, and the payoff matrix is

$$\begin{pmatrix} (-1, 1) & (1, -1) \\ (1, -1) & (-1, 1) \end{pmatrix}. \tag{2.1}$$

If the outcome is $(-1, 1)$, player I (who chooses the row of the payoff matrix) would have done better to choose the other strategy; if the outcome is $(1, -1)$, it is player II, the column player, who would have done better to switch. If players could out-guess each other, they would be trapped in a vicious circle.

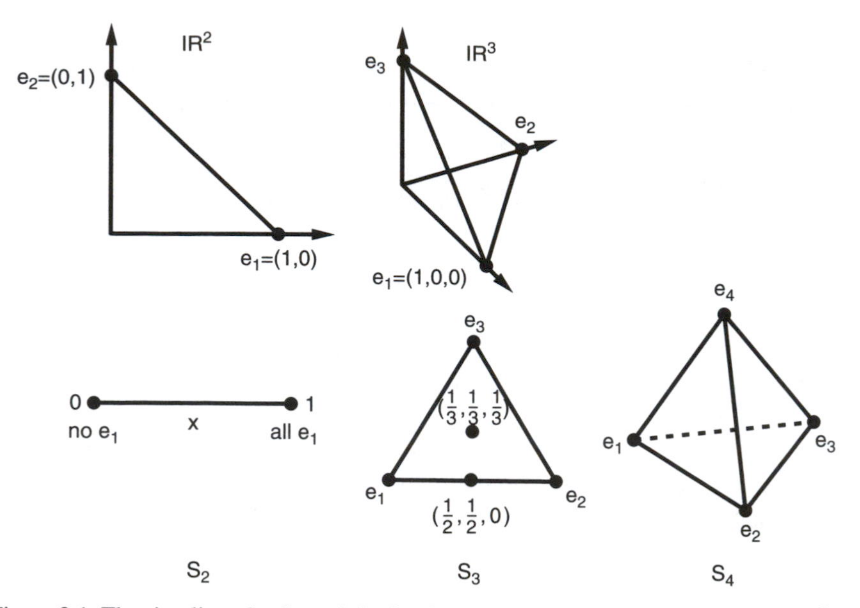

Figure 2.1 The simplices S_2, S_3, and S_4. On the top row, S_2 and S_3 are imbedded in R^2 and R^3 respectively.

2.2 MIXED STRATEGIES

For both players, it is clearly important not to have their decision anticipated by the co-player. A good way to achieve this is to randomize, i.e., to let chance decide. Suppose that player I opts to play strategy $\mathbf{e}_i$ with probability x_i. This *mixed* strategy is thus given by a stochastic vector $\mathbf{x} = (x_1, \ldots, x_n)$ (with $x_i \geq 0$ and $x_1 + \cdots + x_n = 1$). We denote the set of all such mixed strategies by S_n: this is a simplex in R^n, spanned by the unit vectors $\mathbf{e}_i$ of the standard base, which are said to be the *pure* strategies, and correspond to the original set of alternatives, see figure 2.1. (All components of $\mathbf{e}_i$ are 0 except the i-th component, which is 1.)

Similarly, a mixed strategy for player II is an element $\mathbf{y}$ of the unit simplex S_m. If player I uses the pure strategy $\mathbf{e}_i$ and player II uses strategy $\mathbf{y}$, then the payoff for player I (or more precisely, its expected value) is

$$(A\mathbf{y})_i = \sum_{j=1}^{m} a_{ij} y_j. \tag{2.2}$$

If player I uses the mixed strategy $\mathbf{x}$, and II uses $\mathbf{y}$, the payoff for player I is

$$\mathbf{x} \cdot A\mathbf{y} = \sum_i x_i (A\mathbf{y})_i = \sum_{i,j} a_{ij} x_i y_j, \tag{2.3}$$

and the payoff for player II, similarly, is

$$\mathbf{x} \cdot B\mathbf{y} = \sum_{i,j} b_{ij} x_i y_j. \tag{2.4}$$

(The dot on the left hand side denotes the dot product, or Euclidean product, of two vectors.)

If player I knows, by any chance, the strategy $\mathbf{y}$ of the co-player, then player I should use a strategy that is a *best reply* to $\mathbf{y}$. The set of best replies is the set

$$BR(\mathbf{y}) = \arg\max_{\mathbf{x}} \mathbf{x} \cdot A\mathbf{y}, \tag{2.5}$$

i.e., the set of all $\mathbf{x} \in S_n$ such that

$$\mathbf{z} \cdot A\mathbf{y} \leq \mathbf{x} \cdot A\mathbf{y} \tag{2.6}$$

holds for all $\mathbf{z} \in S_n$. Player I has no incentive to deviate from $\mathbf{x}$.

Since the function $\mathbf{z} \mapsto \mathbf{z} \cdot A\mathbf{y}$ is continuous and S_n is compact, the set of best replies is always non-empty. It is a convex set. Moreover, if $\mathbf{x}$ belongs to $BR(\mathbf{y})$, so do all pure strategies in the *support* of $\mathbf{x}$, i.e., all $\mathbf{e}_i$ for which $x_i > 0$. Indeed, for all i,

$$(A\mathbf{y})_i = \mathbf{e}_i \cdot A\mathbf{y} \leq \mathbf{x} \cdot A\mathbf{y}. \tag{2.7}$$

If the inequality sign were strict for some i with $x_i > 0$, then $x_i(A\mathbf{y})_i < x_i(\mathbf{x} \cdot A\mathbf{y})$; summing over all $i = 1, \ldots, n$ then leads to a contradiction. It follows that the set $BR(\mathbf{y})$ is a face of the simplex S_n. It is spanned by the pure strategies which are best replies to $\mathbf{y}$.

2.3 NASH EQUILIBRIUM

If player I has found a best reply to the strategy $\mathbf{y}$ of player II, player I has no reason not to use it—as long as player II sticks to $\mathbf{y}$.

But will player II stick to $\mathbf{y}$? Only if player II has no incentive to use another strategy, i.e., has also hit upon a best reply. Two strategies $\mathbf{x}$ and $\mathbf{y}$ are said to form a *Nash equilibrium* pair if each is a best reply to the other, i.e., if $\mathbf{x} \in BR(\mathbf{y})$ and $\mathbf{y} \in BR(\mathbf{x})$, or alternatively if

$$\mathbf{z} \cdot A\mathbf{y} \leq \mathbf{x} \cdot A\mathbf{y} \tag{2.8}$$

holds for all $\mathbf{z} \in S_n$, and

$$\mathbf{x} \cdot B\mathbf{w} \leq \mathbf{x} \cdot B\mathbf{y} \tag{2.9}$$

holds for all $\mathbf{w} \in S_m$. A Nash equilibrium pair $(\mathbf{x}, \mathbf{y})$ satisfies a minimal consistency requirement: no player has an incentive to deviate (as long as the other player does not deviate either).

A basic result states that Nash equilibrium pairs always exist for any game (A, B). This will be proved in section 2.11. The result holds for vastly wider classes of games than considered so far; it holds for any number of players, any convex compact sets of strategies, any continuous payoff functions, and even beyond. But it would not hold if we had not allowed for mixed strategies: this is shown by the simple example from section 2.1 (betting on even or odd). In that case, the mixed strategies of choosing, with equal probability $1/2$, an even or an odd number, obviously lead to an equilibrium pair: each player gains, on average, zero dollars, and none has an incentive to deviate. On the other hand, if player I uses any other strategy (x_1, x_2) against the $(1/2, 1/2)$ of player II, player I would still have an expected payoff of zero. However, the *other* player would then have an incentive to deviate: whenever $x_1 > x_2$, the best reply for II would be to play $(1, 0)$. If player II did that, however, player I would do better to play $(0, 1)$, and the vicious circle would be in full swing.

In this example, $(\mathbf{x}, \mathbf{y})$ with $\mathbf{x} = \mathbf{y} = (1/2, 1/2)$ is the unique Nash equilibrium pair. We have seen that as long as player II chooses the equilibrium strategy $\mathbf{y}$, player I has no reason to deviate from the equilibrium strategy $\mathbf{x}$; but that on the other hand, player I has no reason not to deviate, either. This would be different if $(\mathbf{x}, \mathbf{y})$ were a *strict* Nash equilibrium pair, i.e., if

$$\mathbf{z} \cdot A\mathbf{y} < \mathbf{x} \cdot A\mathbf{y} \tag{2.10}$$

holds for all $\mathbf{z} \neq \mathbf{x}$, and

$$\mathbf{x} \cdot B\mathbf{w} < \mathbf{x} \cdot B\mathbf{y} \tag{2.11}$$

holds for all $\mathbf{w} \neq \mathbf{y}$. In this case, i.e., when both best-reply sets are singletons, each player will be penalized for unilaterally deviating from the equilibrium.

Whereas every game admits a Nash equilibrium pair, many games admit no strict Nash equilibrium pair; the number game from section 2.1 is an example.

Moreover, even if there exists a strict Nash equilibrium, it can be a let-down, as the Prisoner's Dilemma example from section 1.3 shows. This game has a unique Nash equilibrium, which is strict: both players defect, i.e., $\mathbf{x} = \mathbf{y} = (0, 1)$. Each player, in that case, would be penalized for deviating unilaterally. If both players, however, were to deviate, and opt for cooperation, they would be better off.

A further caveat applies: for many games, there exists not one, but several equilibrium pairs. Which one should the players choose? They could, of course, sit down and talk it over, but this is not considered a solution. In many cases players cannot communicate—sometimes this is prohibited by explicit rules, and sometimes it is just a waste of breath. Consider the Snowdrift game from section 1.4, for instance. In that case, it is easy to see that $(\mathbf{e}_1, \mathbf{f}_2)$ and $(\mathbf{e}_2, \mathbf{f}_1)$ are two Nash equilibrium pairs. They look similar to a bystander, but certainly not to the players themselves. The strategy pair $(\mathbf{e}_1, \mathbf{f}_2)$ means that player I shovels and player II leans back and relaxes. Player I will not like this, but has no incentive not to shovel—for refusing to shovel means spending the night in the car. Of course player I would prefer the other Nash equilibrium pair. But if player I aims at that other equilibrium, and consequently uses strategy $\mathbf{e}_2$, while player II stubbornly clings to the strategy $\mathbf{f}_2$ corresponding to the equilibrium pair which is better for II, then both players end up with the strategy pair $(\mathbf{e}_2, \mathbf{f}_2)$ (an uncomfortably cold night in the car), which is not a Nash equilibrium pair.

2.4 POPULATION GAMES

So far, we have considered games between two specific players trying to guess each other's strategy and find a best reply. Let us now shift perspective, and consider a *population* of players, each with a given strategy. From time to time, two players meet randomly and play the game, using their strategies. We shall consider these strategies as behavioral programs. Such programs can be learned, or inherited, or imprinted in any other way. In a biological setting, strategies correspond to different types of individuals (or behavioral phenotypes).

In order to analyze this set-up, it is convenient to assume that all individuals in the population are indistinguishable, except in their way of interacting, i.e., that the players differ only by their type, or strategy. This applies well to certain games such as the Prisoner's Dilemma, where both players are on an equal footing; but for many other examples of social interactions, there is an inherent asymmetry—for instance, between buyers and sellers.

For simplicity, we start by considering only *symmetric* games. In the case of two-player games, this means that the game remains unchanged if I and II are permuted. In particular, the two players have the same set of strategies. Hence we assume that $n = m$ and $\mathbf{f}_j = \mathbf{e}_j$ for all j; and if a player plays strategy $\mathbf{e}_i$ against someone using strategy $\mathbf{e}_j$ (which is the former $\mathbf{f}_j$), then that player receives the same payoff, whether labeled I or II. Hence $a_{ij} = b_{ji}$, or in other words $B = A^T$ (the transpose of matrix A). Thus a symmetric game is specified by the pair (A, A^T), and therefore is defined by a single, square payoff matrix A.

As we have seen with the Snowdrift example, a symmetric game can have asymmetric Nash equilibrium pairs. These are plainly irrelevant, as long as it is impossible to distinguish players I and II. Of interest are only symmetric Nash equilibrium pairs, i.e., pairs of strategies $(\mathbf{x}, \mathbf{y})$ with $\mathbf{x} = \mathbf{y}$. A *symmetric Nash equilibrium*, thus, is specified by *one* strategy $\mathbf{x}$ having the property that it is a best reply to itself (i.e., $\mathbf{x} \in BR(\mathbf{x})$). In other words, we must have

$$\mathbf{z} \cdot A\mathbf{x} \leq \mathbf{x} \cdot A\mathbf{x} \tag{2.12}$$

for all $\mathbf{z} \in S_n$. A symmetric *strict* Nash equilibrium is accordingly given by the condition

$$\mathbf{z} \cdot A\mathbf{x} < \mathbf{x} \cdot A\mathbf{x} \tag{2.13}$$

for all $\mathbf{z} \neq \mathbf{x}$.

We shall soon prove that every symmetric game admits a symmetric Nash equilibrium.

2.5 SYMMETRIZING A GAME

There is an obvious way to turn a non-symmetric game (A, B) into a symmetric game: simply by letting a coin toss decide who of the two players will be labeled player I. A strategy for this *symmetrized* game must therefore specify what to do in role I, and what in role II, i.e., such a strategy is given by a pair $(\mathbf{e}_i, \mathbf{f}_j)$. A mixed strategy is given by an element $\mathbf{z} = (z_{ij}) \in S_{nm}$, where z_{ij} denotes the probability to play $\mathbf{e}_i$ when in role I and $\mathbf{f}_j$ when in role II. To the probability distribution $\mathbf{z}$ correspond its *marginals*: $x_i = \sum_j z_{ij}$ and $y_j = \sum_i z_{ij}$. The vectors $\mathbf{x} = (x_i)$ and $\mathbf{y} = (y_j)$ belong to S_n and S_m, respectively. It is easy to see that for any given $\mathbf{x} \in S_n$ and $\mathbf{y} \in S_m$ there exists a $\mathbf{z} \in S_{nm}$ having $\mathbf{x}$ and $\mathbf{y}$ as marginals, for instance

$z_{ij} = x_i y_j$ (barring exceptions, there exist many probability distributions with the same marginals).

The payoff for a player using $(\mathbf{e}_i, \mathbf{f}_j)$ against a player using $(\mathbf{e}_k, \mathbf{f}_l)$, with $i, k \in \{1, \dots, n\}$ and $j, l \in \{1, \dots, m\}$, depends on the outcome of the coin-toss and is given by

$$c_{ij,kl} = \frac{1}{2}a_{il} + \frac{1}{2}b_{kj}. \tag{2.14}$$

Let us briefly take for granted that every symmetric game has a symmetric Nash equilibrium. Then it can easily be deduced that every game (A, B) has a Nash equilibrium pair.

Indeed, let us assume that $\bar{\mathbf{z}} \in S_{nm}$ is a symmetric Nash equilibrium for the symmetrized game (C, C^T). This means that

$$\mathbf{z} \cdot C\bar{\mathbf{z}} \leq \bar{\mathbf{z}} \cdot C\bar{\mathbf{z}} \tag{2.15}$$

for all $\mathbf{z} \in S_{nm}$. Let $\mathbf{x}$ and $\mathbf{y}$ be the marginals of $\mathbf{z}$, and $\bar{\mathbf{x}}$ and $\bar{\mathbf{y}}$ the marginals for $\bar{\mathbf{z}}$. Then

$$\mathbf{z} \cdot C\bar{\mathbf{z}} = \sum_{ijkl} z_{ij} c_{ij,kl} \bar{z}_{kl} \tag{2.16}$$

$$= \frac{1}{2}\sum_{ijkl} z_{ij} a_{il} \bar{z}_{kl} + \frac{1}{2}\sum_{ijkl} z_{ij} b_{kj} \bar{z}_{kl} \tag{2.17}$$

$$= \frac{1}{2}\sum_{il} x_i a_{il} \bar{y}_l + \frac{1}{2}\sum_{jk} y_j b_{kj} \bar{x}_k = \frac{1}{2}\mathbf{x} \cdot A\bar{\mathbf{y}} + \frac{1}{2}\mathbf{y} \cdot B^T\bar{\mathbf{x}}. \tag{2.18}$$

Since $\bar{\mathbf{z}}$ is a symmetric Nash equilibrium, equation (2.15) implies

$$\mathbf{x} \cdot A\bar{\mathbf{y}} + \bar{\mathbf{x}} \cdot B\mathbf{y} \leq \bar{\mathbf{x}} \cdot A\bar{\mathbf{y}} + \bar{\mathbf{x}} \cdot B\bar{\mathbf{y}}. \tag{2.19}$$

For $\mathbf{y} = \bar{\mathbf{y}}$ this yields

$$\mathbf{x} \cdot A\bar{\mathbf{y}} \leq \bar{\mathbf{x}} \cdot A\bar{\mathbf{y}}, \tag{2.20}$$

and for $\mathbf{x} = \bar{\mathbf{x}}$,

$$\bar{\mathbf{x}} \cdot B\mathbf{y} \leq \bar{\mathbf{x}} \cdot B\bar{\mathbf{y}}. \tag{2.21}$$

Hence $\bar{\mathbf{x}} \in BR(\bar{\mathbf{y}})$ and $\bar{\mathbf{y}} \in BR(\bar{\mathbf{x}})$, i.e., $(\bar{\mathbf{x}}, \bar{\mathbf{y}})$ is a Nash equilibrium pair of the game (A, B).

2.6 POPULATION DYNAMICS MEETS GAME THEORY

We now consider a symmetric game with payoff matrix A and assume that in a large, well-mixed population, a fraction x_i uses strategy $\mathbf{e}_i$, for $i = 1, \dots, n$. The state of the population is thus given by the vector $\mathbf{x} \in S_n$. A player with strategy $\mathbf{e}_i$ has as expected payoff

$$(A\mathbf{x})_i = \sum_j a_{ij} x_j. \tag{2.22}$$

Indeed, this player meets with probability x_j a co-player using $\mathbf{e}_j$. The average payoff in the population is given by

$$\mathbf{x} \cdot A\mathbf{x} = \sum_i x_i (A\mathbf{x})_i . \tag{2.23}$$

It should be stressed that we are committing an abuse of notation. The same symbol $\mathbf{x} \in S_n$ which denoted in the previous sections the mixed strategy of one specific player now denotes the state of a population consisting of different types, each type playing its pure strategy. (We could also have the players use mixed strategies, but there will be no need to consider this case.)

Now comes an essential step: we shall assume that populations can evolve, in the sense that the frequencies x_i change with time. Thus we let the state $\mathbf{x}(t)$ depend on time, and denote by $\dot{x}_i(t)$ the velocity with which x_i changes, i.e., $\dot{x}_i = dx_i/dt$. In keeping with our population dynamical approach, we shall be particularly interested in the (per capita) growth rates $\dot{x}_i/x_i$ of the frequencies of the strategies.

How do the frequencies of strategies evolve? How do they grow and diminish? There are many possibilities for modeling this process. We shall mostly assume that the state of the population evolves according to the *replicator equation*. This equation holds if the growth rate of a strategy's frequency corresponds to the strategy's payoff, or more precisely to the difference between its payoff $(A\mathbf{x})_i$ and the average payoff $\mathbf{x} \cdot A\mathbf{x}$ in the population. Thus we posit

$$\dot{x}_i = x_i[(A\mathbf{x}_i) - \mathbf{x} \cdot A\mathbf{x}] \tag{2.24}$$

for $i = 1, \dots, n$. Accordingly, a strategy $\mathbf{e}_i$ will spread or dwindle depending on whether it does better or worse than average.

This yields a deterministic model for the state of the population. Indeed, any *ordinary differential equation* $\dot{\mathbf{x}} = \mathbf{F}(\mathbf{x})$ with a smooth right hand side (such as eq. (2.24)) has a unique solution for each initial condition $\mathbf{x}$, i.e., a function $t \mapsto \mathbf{x}(t)$ from an open interval I (containing 0) into R^n such that $\mathbf{x}(0) = \mathbf{x}$ and such that $\dot{\mathbf{x}}(t) = \mathbf{F}(\mathbf{x}(t))$ holds for all $t \in I$. For all differential equations that we consider in this book, the interval I can always be taken to be the whole real line R.

We may interpret the right hand side of the differential equation as a vector field $\mathbf{x} \mapsto \mathbf{F}(\mathbf{x})$. It associates to each point $\mathbf{x}$ in the domain of definition of $\mathbf{F}$ (an open subset B of R^n) the "wind velocity" $\mathbf{F}(\mathbf{x}) \in R^n$ at that point. The solution then describes the motion of a particle, released at time 0 at $\mathbf{x}$ and carried along by the wind. At a point $\mathbf{z}$ such that $\mathbf{F}(\mathbf{z}) = \mathbf{0}$, the velocity is zero. This corresponds to a *rest point*: a particle released at $\mathbf{z}$ will not move. We note that multiplying the right hand side $\mathbf{F}(\mathbf{x})$ by a positive function $M(\mathbf{x}) > 0$ corresponds to a *change in velocity*. The particle will then travel with a different speed, but along the same orbit.

Before we try to explain (in section 2.7) why we are interested in equation (2.24), let us note that $\sum \dot{x}_i = 0$. Furthermore, it is easy to see that the constant function $x_i(t) = 0$ for all t obviously satisfies the i-th equation in (2.24). From this follows that the state space, i.e., the simplex S_n, is invariant: if $\mathbf{x}(0) \in S_n$ then $\mathbf{x}(t) \in S_n$ for all $t \in R$. The same holds for all sub-simplices of S_n, (which are given by $x_i = 0$ for one or several i), and hence also for the boundary bdS_n of S_n (i.e., the union of all such sub-simplices), and moreover also for the interior $int S_n$ of the simplex (the subset satisfying $x_i > 0$ for all i).

2.7 IMITATION DYNAMICS

The replicator equation initially showed up in the context of biological games. The assumption that payoff corresponds to reproductive success, and that individuals breed true, leads almost immediately to this equation. Clearly, for the economic games we are considering here, strategies are unlikely to be inherited, but they can be transmitted through social learning. If we assume that individuals imitate each other, we meet the replicator equation again.

To be more precise, let us assume that from time to time, a randomly chosen individual randomly samples a model from the population and imitates that model with a certain likelihood. Thus the probability that during an interval of length Δt, an individual switches from strategy $\mathbf{e}_j$ to $\mathbf{e}_i$ is given by $x_i f_{ij} \Delta t$. The corresponding input-output equation is

$$x_i(t + \Delta t) - x_i(t) = \sum f_{ij} x_i x_j \Delta t - \sum f_{ji} x_i x_j \Delta t, \tag{2.25}$$

which in the limit $\Delta t \to 0$ yields

$$\dot{x}_i = x_i \sum_j (f_{ij} - f_{ji}) x_j. \tag{2.26}$$

In general, the rates f_{ij} will depend on the state $\mathbf{x}$. For instance, we can assume that

$$f_{ij} = [(A\mathbf{x})_i - (A\mathbf{x})_j]_+. \tag{2.27}$$

This means that an $\mathbf{e}_j$ player comparing himself with an $\mathbf{e}_i$ player will adopt the latter's strategy only if it promises a higher payoff: and if this is the case, the switch is more likely if the difference in payoff is higher. In that case, since $f_{ij} - f_{ji} = (A\mathbf{x})_i - (A\mathbf{x})_j$, the input-output equation yields

$$\dot{x}_i = x_i \sum_j [(A\mathbf{x})_i - (A\mathbf{x})_j] x_j = x_i [(A\mathbf{x})_i - \mathbf{x} \cdot A\mathbf{x}], \tag{2.28}$$

which is just the replicator equation (2.24). We would obtain it in a similar way if, instead of the payoff $(A\mathbf{x})_i$, we use a more general "fitness" term measuring the success of a strategy, for instance $(1 - s)B + s(A\mathbf{x})_i$, with $0 < s \leq 1$. This is the convex combination of a "baseline fitness" $B = B(\mathbf{x}) > 0$ (the same for all types) and the payoff. The size of s specifies the importance of the game in evaluating the "appeal" of a strategy.

We could also assume that

$$f_{ij} = (1 - s)B + s(A\mathbf{x})_i, \tag{2.29}$$

which means that the switching rate depends only on the success of the model (and not on the payoff of the $\mathbf{e}_j$ player); or that

$$f_{ij} = (1 - s)B - s(A\mathbf{x})_j, \tag{2.30}$$

which means that players are all the more prone to imitate one another the more reason they have to be dissatisfied with their own payoff. The role of the convex combination is to guarantee that (at least for small s) the rate is positive.

Not every imitation mechanism leads to the replicator equation. For instance, we could assume that if two players compare their payoffs, the better will always be imitated by the worse. Thus $f_{ij} = 0$ if $(A\mathbf{x})_i < (A\mathbf{x})_j$, $f_{ij} = 1$ if $(A\mathbf{x})_i > (A\mathbf{x})_j$, and $f_{ij} = \frac{1}{2}$, say, in the case of a tie. This leads to a differential equation with a discontinuous right hand side. The dynamics reduces, incidentally, to that of a replicator equation in every region of the state space defined by a specific ordering of the payoff values $(A\mathbf{x})_i$.

Not all learning is social learning (i.e., learning from others). We can also learn from our own experience, for instance by using mostly those strategies that have brought success so far. Moreover, social learning could disregard the success of a model, for instance, by simply imitating whatever is most frequent.

It is worth emphasizing that imitation (like selection, in genetics) does not produce anything new. If a strategy $\mathbf{e}_i$ is absent from the population, it will remain so (i.e., if $x_i(t) = 0$ holds for some time t, it holds for all t). There exist game dynamics that are more innovative. For instance, clever players could adopt the strategy that offers the highest payoff, even if no one in the population is currently using it. Other innovative dynamics arise if we assume a steady rate of switching randomly to other strategies. This can be interpreted as an "exploration rate," and corresponds to a mutation term in genetics.

2.8 BASIC PROPERTIES OF THE REPLICATOR EQUATION

It is easy to see that if we add an arbitrary function $f(\mathbf{x})$ to all payoff terms $(A\mathbf{x})_i$, the replicator equation (2.24) remains unchanged: what is added to the payoff is also added to the average payoff $\mathbf{x} \cdot A\mathbf{x}$, since $\sum x_i = 1$, and cancels out in the difference of the two terms. In particular, this implies that we can add a constant c_j to the j-th column of A (for $j = 1, \ldots, n$) without altering the replicator dynamics in S_n. We shall frequently use this to simplify the analysis.

Another useful property is the quotient rule: if $x_j > 0$, then the time-derivative of the quotient satisfies

$$\left(\frac{x_i}{x_j}\right)^{\cdot} = \left(\frac{x_i}{x_j}\right)[(A\mathbf{x})_i - (A\mathbf{x})_j]. \tag{2.31}$$

More generally, if $V = \prod x_i^{p_i}$ then

$$\dot{V} = V\left[\mathbf{p} \cdot A\mathbf{x} - \left(\sum p_i\right)\mathbf{x} \cdot A\mathbf{x}\right]. \tag{2.32}$$

The rest points $\mathbf{z}$ of the replicator equation are those for which all payoff values $(A\mathbf{z})_i$ are equal, for all indices i for which $z_i > 0$. The common value of these payoffs is the average payoff $\mathbf{z} \cdot A\mathbf{z}$. In particular, all vertices $\mathbf{e}_i$ of the simplex S_n are rest points. (Obviously, if all players are of the same type, imitation leads to no change.)

The replicator equation admits a rest point in $intS_n$ if there exists a solution (in $intS_n$) of the linear equations

$$(A\mathbf{x})_1 = \cdots = (A\mathbf{x})_n. \tag{2.33}$$

Similarly, all rest points on each face can be obtained by solving a corresponding system of linear equations. Typically, each sub-simplex (and S_n itself) contains one or no rest point in its interior.

One can show that if no rest point in S_n exists in the interior of S_n, then all orbits in $intS_n$ converge to the boundary, for $t \to \pm\infty$. In particular, if strategy $\mathbf{e}_i$ is *strictly dominated*, i.e., if there exists a $\mathbf{w} \in S_n$ such that $(A\mathbf{x})_i < \mathbf{w} \cdot A\mathbf{x}$ holds for all $\mathbf{x} \in S_n$, then $x_i(t) \to 0$ for $t \to +\infty$. In the converse direction, if there exists an orbit $\mathbf{x}(t)$ bounded away from the boundary of S_n (i.e., such that for some $a > 0$ the inequality $x_i(t) > a$ holds for all $t > 0$ and all $i = 1, \ldots, n$), then there exists a rest point in $intS_n$. One just has to note that for $i = 1, \ldots, n$,

$$(\log x_i)^{\cdot} = \dot{x}_i / x_i = (A\mathbf{x}(t))_i - \mathbf{x}(t) \cdot A\mathbf{x}(t). \tag{2.34}$$

Integrating this from 0 to T, and dividing by T, leads on the left hand side to $[\log x_i(T) - \log x_i(0)]/T$, which converges to 0 for $T \to +\infty$. The corresponding limit on the right hand side implies that for the accumulation points z_i of the time averages

$$z_i(T) = \frac{1}{T} \int_0^T x_i(t)dt, \tag{2.35}$$

the relations $z_i \geq a > 0$, $\sum z_i = 1$, and

$$\sum a_{1j} z_j = \cdots = \sum a_{nj} z_j \tag{2.36}$$

must hold. Thus $\mathbf{z}$ is a rest point in $intS_n$.

2.9 THE CASE OF TWO STRATEGIES

Let us discuss the replicator equation when there are only two types in the population. Since the equation remains unchanged if we subtract the diagonal term in each column, we can assume without restricting generality that the 2×2 matrix A is of the form

$$\begin{pmatrix} 0 & a \\ b & 0 \end{pmatrix}. \tag{2.37}$$

Since $x_2 = 1 - x_1$, it is enough to observe x_1, which we denote by x. Thus $x_2 = 1 - x$, and

$$\dot{x} = x[(A\mathbf{x})_1 - \mathbf{x} \cdot A\mathbf{x}] = x[(A\mathbf{x})_1 - (x(A\mathbf{x})_1 + (1-x)(A\mathbf{x})_2)], \tag{2.38}$$

and hence

$$\dot{x} = x(1-x)[(A\mathbf{x})_1 - (A\mathbf{x})_2]. \tag{2.39}$$

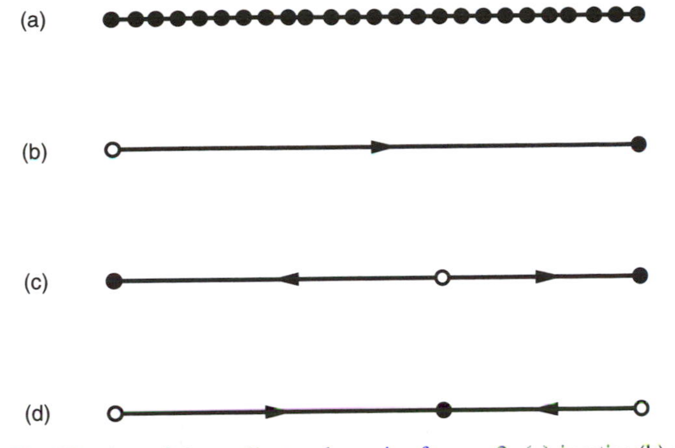

Figure 2.2 Classification of the replicator dynamics for $n = 2$: (a) inertia; (b) dominance; (c) bi-stability; (d) stable coexistence. Circles denote rest points. Filled circles correspond to stable rest points.

Since $(A\mathbf{x})_1 = a(1 - x)$ and $(A\mathbf{x})_2 = bx$, (2.39) reduces to

$$\dot{x} = x(1 - x)[a - (a + b)x]. \tag{2.40}$$

We note that

$$a = \lim_{x \to 0} \frac{\dot{x}}{x}. \tag{2.41}$$

Hence a corresponds to the limit of the per capita growth rate of the missing strategy $\mathbf{e}_1$. Alternatively,

$$a = \frac{d\dot{x}}{dx}, \tag{2.42}$$

where the derivative is evaluated at $x = 0$.

Let us omit the trivial case $a = b = 0$: in this case all points of the state space S_2 (i.e., the interval $0 \le x \le 1$) are rest points. The right hand side of our differential equation is a product of three factors, the first vanishing at 0 and the second at 1; the third factor has a zero $\hat{x} = \frac{a}{a+b}$ in the open interral $]0, 1[$ if and only if $ab > 0$. Thus we obtain three possible cases, see figure 2.2:

1. There is no fixed point in the interior of the state space. This happens if and only if $ab \le 0$. In this case, $\dot{x}$ always has the same sign in $]0, 1[$. If this sign is positive (i.e., if $a \ge 0$ and $b \le 0$, at least one inequality being strict,) this means that $x(t) \to 1$ for $t \to +\infty$, for every initial value $x(0)$ with $0 < x(0) < 1$. The strategy $\mathbf{e}_1$ is said to *dominate* strategy $\mathbf{e}_2$. It is always the best reply, for any value of $x \in]0, 1[$. Conversely, if the sign of $\dot{x}$ is negative, then $x(t) \to 0$ and $\mathbf{e}_2$ dominates. In each case, the dominating strategy converges towards fixation.

As an example, we consider the Prisoner's Dilemma game from section 1.3. The payoff matrix is transformed into

$$\begin{pmatrix} 0 & -5 \\ 5 & 0 \end{pmatrix} \tag{2.43}$$

and defection dominates.

2. There exists a rest point $\hat{x}$ in $]0, 1[$ (i.e., $ab > 0$), and both a and b are negative. In this case $\dot{x} < 0$ for $x \in]0, \hat{x}[$ and $\dot{x} > 0$ for $x \in]\hat{x}, 1[$. This means that the orbits lead away from $\hat{x}$: this rest point is unstable. As in the previous case, one strategy will be eliminated: but the outcome, in this *bistable* case, depends on the initial condition. If x is larger than the threshold $\hat{x}$, it will keep growing; if it is smaller, it will vanish—a positive feedback.

As an example, we can consider the repeated Prisoner's Dilemma from section 1.5. The payoff matrix is transformed into

$$\begin{pmatrix} 0 & -5 \\ -45 & 0 \end{pmatrix} \tag{2.44}$$

and it is best to play TFT if the frequency of TFT-players exceeds 10 percent.

3. There exists a rest point $\hat{x}$ in $]0, 1[$ (i.e., $ab > 0$), and both a and b are positive. In this case $\dot{x} > 0$ for $x \in]0, \hat{x}[$ and $\dot{x} < 0$ for $x \in]\hat{x}, 1[$. This negative feedback means that $x(t)$ converges towards $\hat{x}$, for $t \to +\infty$: the rest point $\hat{x}$ is a stable attractor. No strategy eliminates the other: rather, their frequencies converge towards a *stable coexistence*.

This example can be found in the Snowdrift game from section 1.4. The payoff matrix is transformed into

$$\begin{pmatrix} 0 & 10 \\ 15 & 0 \end{pmatrix} \tag{2.45}$$

and the fixed point corresponds to 40 percent helping and 60 percent shirking.

These three cases (dominance, bi-stability and stable coexistence) will be revisited in the next section. But first, we relate the replicator dynamics to the Nash equilibrium concept.

2.10 NASH EQUILIBRIA AND SATURATED REST POINTS

Let us consider a symmetric $n \times n$ game (A, A^T) with a symmetric Nash equilibrium $\mathbf{z}$. This means that

$$\mathbf{x} \cdot A\mathbf{z} \leq \mathbf{z} \cdot A\mathbf{z} \tag{2.46}$$

for all $\mathbf{x} \in S_n$. With $\mathbf{x} = \mathbf{e}_i$, this implies

$$(A\mathbf{z})_i \leq \mathbf{z} \cdot A\mathbf{z} \tag{2.47}$$

for $i = 1, \ldots, n$. Equality must hold for all i such that $z_i > 0$, as we have seen in section 2.2. Hence $\mathbf{z}$ is a rest point of the replicator dynamics. Moreover, it is a *saturated* rest point: this means by definition that if $z_i = 0$, then

$$(A\mathbf{z})_i - \mathbf{z} \cdot A\mathbf{z} \leq 0. \qquad (2.48)$$

Conversely, every saturated rest point is a Nash equilibrium. The two concepts are equivalent.

Every rest point in $\mathrm{int}S_n$ is trivially saturated; but on the boundary, there may be rest points that are not saturated, as we shall presently see. In that case, there exist strategies not present in the population $\mathbf{z}$, that would do better than average (and better, in fact, than every type that is present). Rest points and Nash equilibria have in common that there exists a c such that $(A\mathbf{z})_i = c$ whenever $z_i > 0$; the additional requirement, for a Nash equilibrium, is that $(A\mathbf{z})_i \leq c$ whenever $z_i = 0$.

Hence every symmetric Nash equilibrium is a rest point, but the converse does not hold. Let us discuss this for the examples from the previous section. It is clear that the fixed points $\hat{x} \in]0, 1[$ are Nash equilibria. In case (1), the dominant strategy is a Nash equilibrium, and the other is not. In case (2), both pure strategies are Nash equilibria. In case (3), none of them is a Nash equilibrium. If you play a bi-stable game, you are well advised to choose the same strategy as your co-player; but in the case of stable coexistence, you should choose the opposite strategy. In both cases, however, the two of you might have different ideas about who plays what.

In the bi-stable case, which of the two pure equilibria, $\mathbf{e}_1$ or $\mathbf{e}_2$, should be chosen? The first idea is: the one with the higher payoff (if it exists). This is said to be the *Pareto-optimal* outcome. In the example given in section 1.7, this is clearly the TFT strategy. The definition of Pareto-optimality depends on the actual payoff values, and is not specified by the replicator dynamics: after adding constants to every column of the payoff matrix, a different strategy may be Pareto-optimal.

The Pareto-optimal solution is not always convincing. Consider for instance the payoff matrix

$$\begin{pmatrix} 2 & -1000 \\ 0 & 1 \end{pmatrix} \qquad (2.49)$$

Clearly, $\mathbf{e}_1$ is Pareto-optimal. But will you play it against an unknown adversary? That player might be a fool, and choose $\mathbf{e}_2$. In that case, you would lose much. Obviously, $\mathbf{e}_2$ is the safer Nash equilibrium. (And on second thought, your co-player may not be a fool, but just suspect that you might be one; or suspect that you might suspect, etc., . . .)

In a bi-stable game

$$\begin{pmatrix} \alpha & \beta \\ \gamma & \delta \end{pmatrix} \qquad (2.50)$$

(with $\alpha > \gamma$ and $\delta > \beta$), the strategy $\mathbf{e}_1$ is said to be *risk-dominant* if it provides the higher payoff against a co-player who is as likely to play $\mathbf{e}_1$ as $\mathbf{e}_2$. This means that $(1/2)(\alpha + \beta) > (1/2)(\gamma + \delta)$, or

$$\gamma - \alpha < \beta - \delta. \qquad (2.51)$$

This condition is invariant with respect to adding constants to each column, and implies for the normalized matrix (2.37) that $a > b$, i.e., (since both values are negative) $\hat{x} < 1/2$. Hence the risk-dominant equilibrium, in a bistable 2×2 game, is the one with the larger basin of attraction.

A handful of results about Nash equilibria and rest points of the replicator dynamics are known as *folk theorem of evolutionary game theory*. For instance, any limit, for $t \to +\infty$, of a solution $\mathbf{x}(t)$ starting in $int S_n$ is a Nash equilibrium; and any stable rest point is a Nash equilibrium. (A rest point $\mathbf{z}$ is said to be stable if for any neighborhood U of $\mathbf{z}$ there exists a neighborhood V of $\mathbf{z}$ such that if $\mathbf{x}(0) \in V$ then $\mathbf{x}(t) \in U$ for all $t \geq 0$.) Both results are obvious consequences of the fact that if $\mathbf{z}$ is not Nash, there exists an i and an ϵ such that $(A\mathbf{x})_i - \mathbf{x} \cdot A\mathbf{x} > \epsilon$ for all $\mathbf{x}$ close to $\mathbf{z}$. In the other direction, if $\mathbf{z}$ is a strict Nash equilibrium, then $\mathbf{z}$ is an asymptotically stable rest point (i.e., not only stable, but in addition *attracting* in the sense that for some neighborhood U of $\mathbf{z}$, $\mathbf{x}(0) \in U$ implies $\mathbf{x}(t) \to \mathbf{z}$ for $t \to +\infty$). The converse statements are generally not valid.

2.11 EXISTENCE OF NASH EQUILIBRIA

In order to prove the existence of a symmetric Nash equilibrium for the symmetric game with $n \times n$ matrix A, i.e., the existence of a saturated rest point for the corresponding replicator equation (2.24), we perturb that equation by adding a small constant term $\epsilon > 0$ to each component of the right hand side. Of course, the relation $\sum \dot{x}_i = 0$ will no longer hold. We compensate this by subtracting the term $n\epsilon$ from each growth rate $(A\mathbf{x})_i - \mathbf{x} \cdot A\mathbf{x}$. Thus we consider

$$\dot{x}_i = x_i[(A\mathbf{x})_i - \mathbf{x} \cdot A\mathbf{x} - n\epsilon] + \epsilon. \tag{2.52}$$

Clearly, $\sum \dot{x}_i = 0$ is satisfied again. On the other hand, if $x_i = 0$, then $\dot{x}_i = \epsilon > 0$. This influx term changes the vector field of the replicator equation: at the boundary of S_n, (which is invariant for the unperturbed replicator equation), the vector field of the perturbed equation points towards the interior.

We shall see presently that (2.52) admits at least one rest point in $int S_n$, which we denote by $\mathbf{z}_\epsilon$. It satisfies

$$(A\mathbf{z}_\epsilon)_i - \mathbf{z}_\epsilon \cdot A\mathbf{z}_\epsilon = \epsilon\left(n - \frac{1}{(\mathbf{z}_\epsilon)_i}\right). \tag{2.53}$$

Let ϵ tend to 0, and let $\mathbf{z}$ be an accumulation point of the $\mathbf{z}_\epsilon$ in S_n. The limit on the left hand side exists, it is $(A\mathbf{z})_i - \mathbf{z} \cdot A\mathbf{z}$. Hence the right hand side also has a limit for $\epsilon \to 0$. This limit is 0 if $z_i > 0$, and it is ≤ 0 if $z_i = 0$. This implies that $\mathbf{z}$ is a saturated rest point of the (unperturbed) replicator equation (2.24), and hence corresponds to a Nash equilibrium.

All that remains to be shown is the existence of a rest point for equation (2.52). Readers who know Brouwer's fixed point theorem will need no proof. All others can find it in the next two sections.

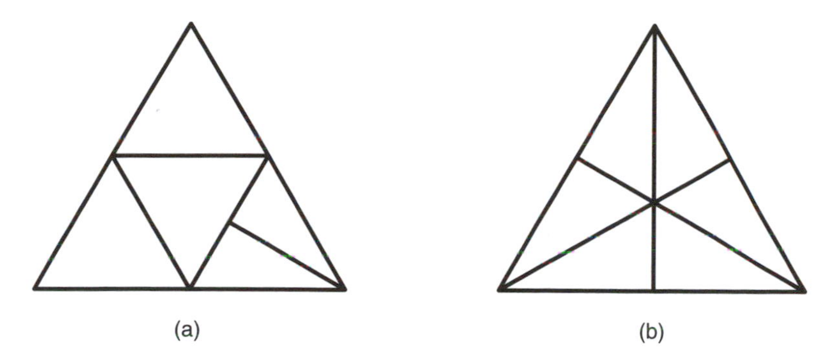

Figure 2.3 (a) This is not a simplicial decomposition. (b) The barycentric decomposition is a simplicial decomposition.

2.12 SPERNER'S LEMMA

Let us consider an $n-1$-dimensional simplex S, i.e., the closed convex hull of n points $\mathbf{y}_1, \ldots, \mathbf{y}_n$ such that the vectors $\mathbf{y}_i - \mathbf{y}_n$, $i = 1, \ldots, n-1$, are linearly independent. Any non-trivial subset of the vertices $\mathbf{y}_1, \ldots, \mathbf{y}_n$ spans a *sub-simplex* of S. The boundary of S is the union of the n *full* (i.e., $n-2$-dimensional) faces. A *simplicial decomposition* of S consists of finitely many $n-1$-dimensional simplices whose union is S and whose interiors are pairwise disjoint. We furthermore require that if two such (closed) sub-simplices are not disjoint, they must share a face: i.e., if the intersection contains a vertex of one sub-simplex, that point is also a vertex of the other, see figure 2.3.

An example is the barycentric subdivision. (The *barycenter* of the simplex S is $(\mathbf{y}_1 + \cdots + \mathbf{y}_n)/n$). We begin with the barycenters of all 1-dimensional sub-simplices, i.e., the midpoints of the edges. They divide the edges of S into 1-dimensional subsimplices. We then introduce the barycenters of the 2-dimensional sub-simplices of S, and consider the 2-dimensional sub-simplices obtained as a convex hull of such a barycenter and a 1-dimensional sub-simplex on the boundary of the corresponding face; and so on into higher dimensions.

Now suppose that we are given a *coloring* of the vertices of the simplicial decomposition by n colors, in the sense that we associate to each vertex an $i \in \{1, \ldots, n\}$. We require that the vertices $\mathbf{y}_i$ of S are colored by the colors i, and that for any sub-simplex of S, only the colors of the vertices spanning that sub-simplex are used. We say that a sub-simplex is I-colored if $I \subset \{1, \ldots, n\}$ is the list of all colors actually occurring at the vertices of that sub-simplex.

Sperner's lemma states that there always exists an odd number of $\{1, \ldots, n\}$-colored sub-simplices. (In particular, we need the full set of colors for at least one sub-simplex of S.)

The proof goes by induction. For $n = 2$ (i.e., for the segment S_2) the statement is obvious. Suppose it is proved up to $n-1$. We can apply this to the boundary face of S, which is opposed to $\mathbf{y}_n$: its simplicial decomposition contains an odd number of sub-simplices which are $\{1, \ldots, n-1\}$-colored.

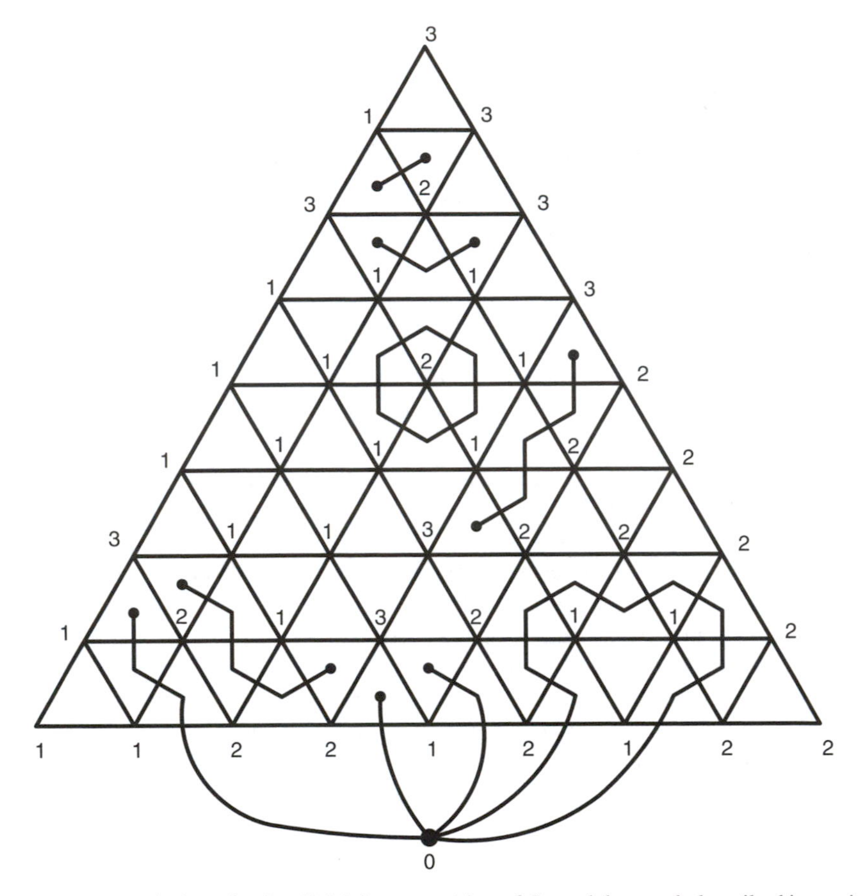

Figure 2.4 A coloring of a simplicial decomposition of S_3, and the graph described in section 2.12.

We now construct a graph having as vertices the barycenters of the sub-simplices of S. We join two such barycenters by an edge if and only if the corresponding sub-simplices share a $\{1, \ldots, n-1\}$-colored face, see figure 2.4. We add one further vertex o lying outside of S, and connect it with the barycenters of those sub-simplices having a $\{1, \ldots, n-1\}$-colored face on the boundary of S. We see immediately that o is connected to an odd number of barycenters, which belong to sub-simplices having a full face belonging to the face of S opposite of $\mathbf{y}_n$.

If a sub-simplex is $\{1, \ldots, n\}$-colored, it has exactly one $\{1, \ldots, n-1\}$-colored face. Hence its barycenter lies on exactly one edge of the graph; it is an end-point of the graph. As to the other barycenters, they either lie on two edges, or on none at all. Indeed, if a sub-simplex that is not $\{1, \ldots, n\}$-colored has an $\{1, \ldots, n-1\}$-colored face, then the opposite vertex must have one of the colors $1, \ldots, n-1$, and hence that sub-simplex has exactly one additional $\{1, \ldots, n-1\}$-colored face.

We note that it is possible that the graph has closed loops. But since an odd number of edges issues from o, there must be an odd number of end-points of the graph, and hence an odd number of fully colored sub-simplices of S. Thus Sperner's lemma holds.

2.13 A FIXED-POINT THEOREM

We now show that a smooth vector field on the plane $\sum x_i = 1$ satisfying

$$\sum \dot{x}_i = 0 \tag{2.54}$$

and

$$x_i = 0 \Rightarrow \dot{x}_i > 0 \tag{2.55}$$

has a fixed point in $int S_n$. We proceed indirectly and assume that it has no fixed point. To each point $\mathbf{x} \in S_n$ we can associate the "color" $i := \min\{j : \dot{x}_j < 0\}$. This is possible because $\dot{\mathbf{x}} \neq \mathbf{0}$ and $\sum_j \dot{x}_j = 0$. We note that condition (2.55) implies that on each sub-simplex of S_n, only the colors of the vertices spanning that face are used.

This induces a coloring for any simplicial decomposition. Each such decomposition must have an odd number of fully-colored sub-simplices. Now consider a sequence of simplicial decompositions whose width (the size of the largest sub-simplex) converges to 0. (For instance, we can start with the barycentric subdivision of S, and then iterate this ad lib.)

This yields a sequence of fully-colored sub-simplices: by compactness, the sub-sequence converges to a point $\mathbf{x} \in S_n$. For each i, this point is a limit of i-colored vertices, and hence must satisfy $\dot{x}_i \leq 0$. Since $\sum \dot{x}_i = 0$ this implies $\dot{\mathbf{x}} = \mathbf{0}$, a contradiction.

Hence the vector field (2.52) must have some fixed point in S_n. This closes the gap in the proof that each replicator equation admits a saturated fixed point.

2.14 ROCK-SCISSORS-PAPER

Whereas there exist only four possible types of replicator dynamics for $n = 2$, there exist about a hundred of them for $n = 3$ (and for $n > 3$ a full classification seems presently out of sight). A particularly interesting example occurs if the three strategies dominate each other in a cyclic fashion, i.e., if $\mathbf{e}_1$ dominates $\mathbf{e}_2$, in the absence of $\mathbf{e}_3$; and similarly if $\mathbf{e}_2$ dominates $\mathbf{e}_3$; and $\mathbf{e}_3$, in turn, dominates $\mathbf{e}_1$. Such a cycle occurs in the game of Rock-Scissors-Paper. If we assume that the winner receives one dollar from the loser, the payoff matrix is

$$\begin{pmatrix} 0 & 1 & -1 \\ -1 & 0 & 1 \\ 1 & -1 & 0 \end{pmatrix}. \tag{2.56}$$

This is a zero-sum game: one player receives what the other player loses. Hence the average payoff in the population, $\mathbf{x} \cdot A\mathbf{x}$, is zero. There exist only four rest points, one in the center, $\mathbf{m} = (1/3, 1/3, 1/3) \in int S_3$, and the other three at the vertices $\mathbf{e}_i$. The only Nash equilibrium is $\mathbf{m}$.

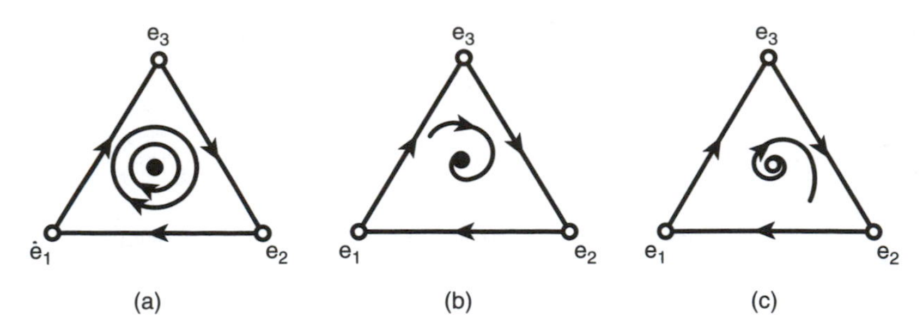

Figure 2.5 The replicator dynamics of the Rock-Scissors-Paper game with payoff matrix (2.58): (a) $a = 1$; (b) $a > 1$; (c) $0 < a < 1$.

Let us consider the function $V := x_1 x_2 x_3$, which is positive in the interior of S_3 (with its maximum at $\mathbf{m}$) and vanishes on the boundary. Using (2.32) we see that $t \to V(\mathbf{x}(t))$ satisfies

$$\dot{V} = V(x_2 - x_3 + x_3 - x_1 + x_1 - x_2) = 0. \tag{2.57}$$

Hence V is a *constant of motion*: all orbits $t \to \mathbf{x}(t)$ of the replicator equation remain on constant level sets of V, see figure 2.5. This implies that all orbits in $int S_n$ are closed orbits surrounding $\mathbf{m}$. The invariant set consisting of the three vertices $\mathbf{e}_i$ and the orbits connecting them along the edges of S_3 is said to form a *heteroclinic set*. Any two points on it can be connected by "shadowing the dynamics." This means to travel along the orbits of that set and, at appropriate times that can be arbitrarily rare, to make an arbitrarily small step. In the present case, it means for instance to flow along an edge towards $\mathbf{e}_1$, and then step onto the edge leading away from $\mathbf{e}_1$. This step can be arbitrarily small: travelers just have to wait until they are sufficiently close to the "junction" $\mathbf{e}_1$.

Now let us consider the game with matrix

$$\begin{pmatrix} 0 & a & -1 \\ -1 & 0 & a \\ a & -1 & 0 \end{pmatrix}. \tag{2.58}$$

For $a > 0$, it has the same structure of cyclic dominance. For $a \neq 1$ the game is no longer a zero sum game, but it has the same rest points. The point $\mathbf{m}$ is a Nash equilibrium and the boundary of S_3 is a heteroclinic set, as before. But now,

$$\mathbf{x} \cdot A\mathbf{x} = (a - 1)(x_1 x_2 + x_2 x_3 + x_3 x_1) \tag{2.59}$$

and hence

$$\dot{V} = V(a - 1)[1 - 3(x_1 x_2 + x_2 x_3 + x_3 x_1)] \tag{2.60}$$

$$= \frac{V(a - 1)}{2}[(x_1 - x_2)^2 + (x_2 - x_3)^2 + (x_3 - x_1)^2]. \tag{2.61}$$

This expression vanishes on the boundary of S_3 and at $\mathbf{m}$. It has the sign of $a - 1$ everywhere else on S_3. If $a > 1$, this means that all orbits cross the constant-level sets of V in the uphill direction, and hence converge to $\mathbf{m}$. This implies that ultimately, all three types will be present in the population in equal frequencies: the rest point $\mathbf{m}$ is asymptotically stable, see figure 2.5b. But for $a < 1$, the orbits flow downhill, towards the boundary of S_3. The Nash equilibrium $\mathbf{m}$ corresponds to an unstable rest point, and the heteroclinic cycle on the boundary attracts all other orbits, see figure 2.5c.

Let us follow the state $\mathbf{x}(t)$ of the population, for $a < 1$. If the state is very close to a vertex, for instance $\mathbf{e}_1$, it is close to a rest point and hence almost at rest. For a long time, the state does not seem to change. Then, it picks up speed and moves towards the vicinity of the vertex $\mathbf{e}_3$, where it slows down and remains for a much longer time, etc. This looks like a recurrent form of "punctuated equilibrium": long periods of quasi-rest followed by abrupt upheavals.

2.15 STOCHASTIC PROCESSES AND FIXATION PROBABILITIES

So far, we have considered the limiting case of infinitely large populations. If we assume a population of finite size, we can no longer rely on deterministic models. In finite populations, random fluctuations, due for instance to sampling effects, have to be taken into account. Instead of ordinary differential equations, we must use stochastic processes.

Let us assume, in the simplest case, that a population of finite size M consists of two types of players only, $\mathbf{e}_1$ and $\mathbf{e}_2$. From time to time, one of the players updates strategy, by imitating a model chosen from the population. The state of the population is given by the number i of individuals of type $\mathbf{e}_1$ (while the number of players of type $\mathbf{e}_2$ is $M - i$). Let p_{ij} be the probability that the transition leads from i to j. The matrix P is tri-diagonal, i.e., $p_{ij} = 0$ if $|j - i| > 1$. The states 0 and M are absorbing: if all individuals are of the same type, imitation can not introduce the other type. We write $p_{i,i+1} = b_i$ and $p_{i,i-1} = d_i$ (because these transition probabilities correspond, in another interpretation, to *birth* and *death* rates).

We denote by p_i the probability that a population in state i will eventually reach state M, i.e., consist entirely of type $\mathbf{e}_1$. This state M is *absorbing*, since once reached it will not be left. The probability p_i that, starting in state i, such a *fixation* of the type $\mathbf{e}_1$ occurs, must satisfy

$$p_i = d_i p_{i-1} + (1 - b_i - d_i) p_i + b_i p_{i+1} \tag{2.62}$$

for $i = 1, \ldots, M-1$. Indeed, in the first updating event, the number of $\mathbf{e}_1$'s will either increase or decrease by 1, or remain unchanged (when a player imitates someone of his own kind); and after this first step, fixation must occur. Moreover, we have $p_0 = 0$ and $p_M = 1$. Setting $y_i := p_i - p_{i-1}$, equation (2.62) can be written as

$$y_{i+1} = \frac{d_i}{b_i} y_i. \tag{2.63}$$

Since $p_1 = y_1$ and $\sum_{i=0}^{k} y_i = p_k - p_0 = p_k$, we obtain

$$1 = p_M = \sum_{i=1}^{M} y_i = p_1 \left(1 + \frac{d_1}{b_1} + \cdots + \frac{d_1}{b_1}\frac{d_2}{b_2} \cdots \frac{d_{M-1}}{b_{M-1}}\right), \tag{2.64}$$

so that

$$p_i = \frac{1 + \sum_{j=1}^{i-1} \prod_{k=1}^{j} d_k/b_k}{1 + \sum_{j=1}^{M-1} \prod_{k=1}^{j} d_k/b_k} \tag{2.65}$$

for $i = 1, \ldots, M$. In particular, we denote by $\rho_{1,2}$ the *fixation probability* p_1, i.e., the probability that a single individual of type $\mathbf{e}_1$ in a population consisting otherwise of type $\mathbf{e}_2$ will eventually be imitated by everyone. It is given by

$$\rho_{1,2} = \frac{1}{1 + \sum_{j=1}^{M-1} \prod_{k=1}^{j} d_k/b_k}. \tag{2.66}$$

So far, we have not specified the imitation mechanism. In this chapter, we shall consider only the so-called Moran process, developed in the context of population genetics. For this reason, we shall adopt the corresponding terminology, and assume that each individual has a certain "fitness," which in our context means some measure of success, such that individuals with a higher fitness are more likely to be imitated (see section 2.7). The Moran process consists in drawing one individual at random (each has the same probability $1/M$ of being chosen) and endowing it with the type of a "model player" who is selected from the population with a probability proportional to that model's success.

Thus let us assume, as a first example, that individuals of type $\mathbf{e}_1$ have fitness r, while those of type $\mathbf{e}_2$ have a fitness normalized to be equal to 1. We then obtain for the death rate

$$d_i = \left(\frac{i}{M}\right)\left(\frac{M-i}{ri + M - i}\right), \tag{2.67}$$

where the first fraction is the probability that the updating individual is of type $\mathbf{e}_1$, and the second that the selected model is of type $\mathbf{e}_2$. Similarly, for the birth rate,

$$b_i = \left(\frac{M-i}{M}\right)\left(\frac{ri}{ri + M - i}\right). \tag{2.68}$$

Hence $d_i/b_i = 1/r$ and

$$\rho_{1,2} = \frac{1 - r^{-1}}{1 - r^{-M}}. \tag{2.69}$$

If $r \to 1$, we obtain as limiting value $\rho_{1,2} = 1/M$, which is reassuring. This is the fixation probability of a *neutral* type, i.e., the probability that a single individual of type $\mathbf{e}_1$, doing exactly as well as the resident $\mathbf{e}_2$ individuals, will eventually be copied by the entire population.

2.16 GAMES IN FINITE POPULATIONS

Now suppose that in a population of size M, individuals are engaged in pairwise games, and strategies are determined by type $\mathbf{e}_1$ or $\mathbf{e}_2$. If the payoff matrix is

$$\begin{pmatrix} \alpha & \beta \\ \gamma & \delta \end{pmatrix}, \tag{2.70}$$

then the expected payoff depends on the state i of the population. For a player of type $\mathbf{e}_1$, it is given by

$$F_i = \alpha \frac{i-1}{M-1} + \beta \frac{M-i}{M-1}, \tag{2.71}$$

and for a player of type $\mathbf{e}_2$ by

$$G_i = \gamma \frac{i}{M-1} + \delta \frac{M-i-1}{M-1}. \tag{2.72}$$

(Players do not play against themselves.) As in section 2.7, we assume that the fitness, i.e., the likelihood to be imitated, is given as a convex combination of the payoff and a "baseline fitness," the same for all, which we normalize to 1. Hence the fitness of an $\mathbf{e}_1$ individual, if the population is in state i, is given by

$$f_i = (1-s) + sF_i \tag{2.73}$$

and that of an $\mathbf{e}_2$ individual is

$$g_i = (1-s) + sG_i. \tag{2.74}$$

Here the parameter $s \in [0, 1]$ measures the "strength of selection," i.e., the importance of the game for overall success. If $s = 0$ the game is irrelevant. In the limiting case of an infinitely large population, the Moran process leads to the switching rate given by equation (2.29) and hence to the replicator dynamics.

The birth and death rates are

$$b_i = \left(\frac{M-i}{M}\right)\left(\frac{if_i}{if_i + (M-i)g_i}\right) \tag{2.75}$$

and

$$d_i = \left(\frac{i}{M}\right)\left(\frac{(M-i)g_i}{if_i + (M-i)g_i}\right), \tag{2.76}$$

so that

$$\frac{d_i}{b_i} = \frac{g_i}{f_i} = \frac{1-s(1-G_i)}{1-s(1-F_i)}. \tag{2.77}$$

The fixation probability (2.66) is therefore given by

$$\rho_{1,2} = \left[1 + \sum_{j=1}^{M-1} \prod_{i=1}^{j} \frac{1-s+sG_i}{1-s+sF_i}\right]^{-1}. \tag{2.78}$$

2.17 LIMITING CASES

For small s, expression (2.77) can be approximated, up to first order, by

$$\frac{d_i}{b_i} = 1 - s(F_i - G_i). \tag{2.79}$$

Now by equations (2.71) and (2.72),

$$H_i := F_i - G_i = \frac{1}{M-1}[\bar{e} + \bar{f}i] \tag{2.80}$$

with $\bar{e} = -\alpha + \beta M - \delta M + \delta$ and $\bar{f} = \alpha - \beta - \gamma + \delta$. Hence up to first order in s, the fixation probability of type $\mathbf{e}_1$ is, according to equation (2.78), given by

$$\rho_{1,2} = \left[1 + \sum_{k=1}^{M-1} \prod_{i=1}^{k} (1 - sH_i)\right]^{-1}. \tag{2.81}$$

It is easy to see that

$$\sum_{k=1}^{M-1} \prod_{i=1}^{k} (1 - sH_i) = M - 1 - s \sum_{i=1}^{M-1} (M-i)H_i, \tag{2.82}$$

and that

$$\sum_{i=1}^{M-1} (M-i)(\bar{e} + \bar{f}i) = M(M-1)\bar{e} + (M\bar{f} - \bar{e}) \sum_{i=1}^{M-1} i - \bar{f} \sum_{1}^{M-1} i^2. \tag{2.83}$$

The first sum on the right hand side is $M(M-1)/2$ and the second sum is $M(M-1)(2M-1)/6$. This yields altogether

$$M(M-1)(M\bar{f} + \bar{f} + 3\bar{e})/6 = M(M-1)(eM - f)/6, \tag{2.84}$$

with $e = \alpha + 2\beta - \gamma - 2\delta$ and $f = 2\alpha + \beta + \gamma - 4\delta$. Up to first order in s, equation (2.81) yields

$$\rho_{1,2} = \left[1 - \frac{s}{6}(eM - f)\right]^{-1} / M. \tag{2.85}$$

We say that strategy $\mathbf{e}_1$ is *advantageous* if its fixation probability is higher than that of a neutral mutant, i.e., if $\rho_{1,2} > 1/M$. This condition reads $eM > f$, i.e.,

$$\alpha(M-2) + \beta(2M-1) > \gamma(M+1) + \delta(2M-4). \tag{2.86}$$

For the limit $M \to \infty$ we obtain

$$\alpha + 2\beta > \gamma + 2\delta, \tag{2.87}$$

or, with the normalization from matrix (2.37), $b < 2a$. This inequality always holds if $\mathbf{e}_1$ dominates $\mathbf{e}_2$, i.e., if $b \leq 0$ and $a \geq 0$ (one inequality being strict). The dominant strategy is always advantageous. In the case of stable coexistence, i.e., $a > 0$ and $b > 0$, it means that $\hat{x} > 1/3$, where $\hat{x} = \frac{a}{a+b}$ is the Nash equilibrium. Thus if $1/3 < \hat{x} < 2/3$, both strategies are advantageous. Finally, in the case of a bi-stable game, i.e., if $a < 0$ and $b < 0$, inequality (2.86) means that $\hat{x} < 1/3$, where $\hat{x}$ is the unstable Nash equilibrium in $]0, 1[$. This means that for the replicator equation, the basin of attraction of $\mathbf{e}_1$ is more than twice as large as that of $\mathbf{e}_2$. In particular, if $\mathbf{e}_1$ is advantageous, it is risk-dominant. If $\hat{x}$ lies between $1/3$ and $2/3$, none of the two strategies is advantageous.

For the examples of a Repeated Prisoner's Dilemma game (section 1.7) or a Snowdrift game (section 1.4), we see from inequality (2.86) that cooperation is advantageous for $M > 4$ resp. $M > 20$.

The fixation probability $\rho_{2,1}$ of $\mathbf{e}_2$ is obtained similarly to that of $\mathbf{e}_1$ (by replacing e with $-2\alpha - \beta + 2\gamma + \delta$ and f with $-4\alpha + \beta + \gamma + 2\delta$). The condition $\rho_{1,2} > \rho_{2,1}$ means

$$(M - 2)(\alpha - \delta) > M(\gamma - \beta). \tag{2.88}$$

In the limit of large M, this reduces to the condition $\alpha - \delta > \gamma - \beta$. This is just the condition $a > b$ that $\mathbf{e}_1$ is risk-dominant.

For any value of $\pi \in [0, 1]$, the vector $(\pi, 0, \dots, 0, 1 - \pi) \in S_M$ is a stationary distribution of the imitation process. Let us assume that with some probability $\mu > 0$, players can change their strategy without imitating another player, just by random trial. In that case, the resulting Markov chain is recurrent. It describes the interplay between innovation and imitation. Let us assume that μ is so small that we can separate the time scales of the two processes. This means that most of the time, the population is in the homogeneous state 0 or M. Occasionally, a single individual tries the other strategy. Then, the imitation process starts anew, leading either to the extinction of the new type or to its fixation. In this "adiabatic" case, the resulting process can be approximated by a Markov chain with two states, 1 and 2, (which correspond to homogeneous populations consisting of type $\mathbf{e}_1$ or $\mathbf{e}_2$). This Markov chain is given by the matrix

$$\begin{pmatrix} 1 - \rho_{2,1} & \rho_{2,1} \\ \rho_{1,2} & 1 - \rho_{1,2} \end{pmatrix} \tag{2.89}$$

whose unique stationary distribution, the left eigenvector

$$\left(\frac{\rho_{1,2}}{\rho_{1,2} + \rho_{2,1}}, \frac{\rho_{2,1}}{\rho_{1,2} + \rho_{2,1}} \right) \tag{2.90}$$

describes the prevalence of the two types, for large time spans. In particular, for the bi-stable case and large M, strategy $\mathbf{e}_1$ is risk-dominant if and only if the stochastic process spends more time in the corresponding homogeneous state.

The same "adiabatic" argument holds also for n types $\mathbf{e}_i$. If the "innovation rate" μ is sufficiently small, the population will always consist of one or at most two types only. If in a homogeneous population, a single individual switches to a different type, then the imitation process will have caused the fixation or the elimination of that type before the next innovation occurs. If we assume that these innovations are random explorations, i.e., that every non-resident type has the same chance $1/(n - 1)$ to occur, we obtain an $n \times n$ Markov chain P with transition probabilities p_{ij} given by

$$p_{ij} = \rho_{j,i}/(n - 1) \tag{2.91}$$

for $i \neq j$. Here, $\rho_{j,i}$ is the fixation probability of j in i, i.e., the probability that a single individual of type j in a population consisting otherwise of type i will eventually be copied by the entire population.

2.18 REFERENCES

The first, and already classic book on evolutionary game theory is by Maynard Smith (1982). Among the textbooks which have appeared since, Weibull (1995), Hofbauer and Sigmund (1998), and Cressman (2003) are closest in spirit to the approach presented here, but see also Colman (1995), Fudenberg and Levine (1998), Gintis (2000), and Vincent and Brown (2005). Hofbauer and Sigmund (2003) and Sandholm (2009) consider more general types of game dynamics. The replicator equation was introduced by Taylor and Jonker (1978), see also Hofbauer, Schuster, and Sigmund (1979) and Zeeman (1980); the name was suggested in Schuster and Sigmund (1983). Derivations via social learning were proposed by Helbing (1992) and Schlag (1997). Bomze (1983) gives a classification of replicator dynamics with three strategies. Hofbauer (2000) presents an account of the relations between the concept of Nash equilibrium pairs, existence proofs, and diverse game dynamics. Our treatment of finite population games closely follows Nowak (2006a), see also Nowak et al. (2004), and Taylor et al. (2004). A different, less "stochastic" but related approach is used in Kandori, Mailath, and Rob (1993) and in Peyton Young and Foster (1995).

Chapter Three

Direct Reciprocity: The Role of Repetition

3.1 HELP

As Darwin wrote, "The small strength and speed of man, his want of natural weapons, etc., are more than counterbalanced . . . by his social qualities which lead him *to give and receive* aid from his fellow-men" (italics added). In its simplest form, to help means to confer a benefit b to another individual, at a cost c to oneself. This can be viewed as an atom of social interaction.

In the *Donation game*, two players have to decide simultaneously (more precisely, in ignorance of the co-player's decision) whether to give help to their co-player or not. The two strategies $\mathbf{e}_1$ and $\mathbf{e}_2$ will be denoted by C (for *cooperate*) and D (for *defect*), respectively. This yields the following payoff matrix:

$$\begin{pmatrix} b-c & -c \\ b & 0 \end{pmatrix}. \tag{3.1}$$

If not otherwise stated, we will assume $b > c > 0$. The second strategy D dominates the first. This is an example of a Prisoner's Dilemma game, as described in section 1.3, i.e., a symmetric 2×2 game whose payoff matrix

$$\begin{pmatrix} R & S \\ T & P \end{pmatrix} \tag{3.2}$$

satisfies

$$T > R > P > S. \tag{3.3}$$

The Prisoner's Dilemma game encapsulates the tug-of-war between the common interest (R is larger than P) and the selfish interest (D dominates C). Selfishness ought to win in this conflict. Indeed, the game has a unique Nash equilibrium, namely defection; and imitation of successful individuals leads inexorably to the demise of cooperation, see section 2.10.

It can be interesting to compare this Donation game with the Snowdrift game (see section 1.4). Both players can receive a benefit b each, if they come up with a fee $c < b$. They have to decide simultaneously whether to pay the fee or not, knowing that if both decide to pay, they will share the cost. The payoff matrix is

$$\begin{pmatrix} b-\frac{c}{2} & b-c \\ b & 0 \end{pmatrix}. \tag{3.4}$$

Obviously, it is best to do the opposite of what the other player does. If your co-player is willing to pay the fee, you yourself can safely skip it. But if your co-player is unwilling to pay the fee, you should better pay. Clearly, a player would prefer to be the one who does not pay the cost. The game has a unique symmetric Nash equilibrium. It consists in paying the fee with a probability of $\frac{2(b-c)}{2b-c} = 1 - \frac{c}{2b-c}$. We note that if the Donation game is played twice, then the two players would do twice as well to both play C each time than to take turns in playing C. In two turns of the Snowdrift game, they would do as well to both play C each time than to take turns in playing C. The Snowdrift game is an example of the so-called Chicken game, a symmetric 2×2 game whose payoff matrix (3.2) satisfies

$$T > R > S > P. \tag{3.5}$$

The small difference in rank order (S and P are permuted) has a considerable effect.

3.2 ITERATED GAMES

Let us now consider several rounds of the simultaneous Donation game. If the number of rounds is known to both players, then backward induction predicts, as seen in section 1.5, that selfish players ought to play D in each round.

Let us suppose instead that the two players do *not* know how many rounds their game will last. Usually, one assumes that after every round, a further round can occur with a constant probability $w < 1$. (One could alternatively assume that the number of rounds is given by a Poisson distribution, for instance.) We number the initial round by 0, and by n the round obtained at the n-th iteration. The probability that the game is iterated at least n times is given by w^n. The probability that the game has *exactly* $n + 1$ rounds (the initial round followed by n iterations) is $w^n(1 - w)$. The number of rounds is a random variable with a geometric distribution, and its expected value is

$$1(1 - w) + 2w(1 - w) + \cdots + nw^{n-1}(1 - w) + \cdots = \frac{1}{1 - w}. \tag{3.6}$$

Let us denote by $A(n)$ the payoff in the n-th round. The expected value of the total payoff is given by

$$\sum_{n=0}^{+\infty} w^n(1 - w)[A(0) + \cdots + A(n)], \tag{3.7}$$

which by *Abel's summation formula* is the power series $A(0) + wA(1) + \cdots$. Since $A(n) \in \{R, S, T, P\}$, all $A(n)$ are uniformly bounded, and hence expression (3.7) always converges to some value $A(w)$, for $0 \leq w < 1$. The average payoff *per round* is given by

$$(1 - w)A(w) = (1 - w) \sum_{n=0}^{+\infty} w^n A(n). \tag{3.8}$$

It is often instructive to analyze the limiting case $w = 1$. In this case, the game consists of infinitely many rounds, and the total payoff $A(0) + A(1) + \cdots$ may diverge. It is convenient, in that case, to consider the average (over time) of the payoff *per round*, namely

$$\lim_{n\to+\infty} \frac{A(0)+\cdots+A(n)}{n+1}, \tag{3.9}$$

provided this limit exists. The *theorem of Frobenius* implies that in this case, expression (3.9) is given by the limit of equation (3.8), i.e., by $\lim_{w\to 1}(1-w)A(w)$.

3.3 THE GOOD, THE BAD, AND THE RECIPROCATOR

Let us first consider the interaction of three strategies only. The cooperator always decides to help; the defector always refuses to help; and the reciprocator refuses to help if and only if the co-player refused to help in the previous round. (By default, thus, the reciprocator donates in the initial round.) These are the strategies $\mathbf{e}_1 = AllC$, $\mathbf{e}_2 = AllD$ and $\mathbf{e}_3 = TFT$, respectively.

We consider a large, well-mixed population. The frequencies of the three strategies are given by x, y, and z, respectively (with $x+y+z=1$). With P_x, P_y, and P_z we denote the expected values for the total payoff obtained by players using these strategies (rather than by $(A\mathbf{x})_1$, etc., as in the previous chapter). The average payoff in the population is $\bar{P} = xP_x + yP_y + zP_z$. We shall assume that more successful strategies are more likely to be imitated, as in section 2.7. Hence the evolution of the frequencies of the three strategies in the population is given by the replicator equation

$$\begin{aligned} \dot{x} &= x(P_x - \bar{P}) \\ \dot{y} &= y(P_y - \bar{P}) \\ \dot{z} &= z(P_z - \bar{P}). \end{aligned} \tag{3.10}$$

We will frequently use the fact that the replicator equation remains unchanged (on the simplex S_3) if the same function is added to each payoff term (see section 2.8), and by abuse of notation still design the corresponding terms with P_x, P_y, P_z, and $\bar{P}$. In particular, we can normalize the payoff matrix by adding an appropriate constant to each column.

AllD against *AllD* obtains payoff $A(n)=0$ in every round, so that $A(w)=0$. *TFT* against *AllD* earns $A(0)=-c$ in the initial round, and henceforth $A(n)=0$ for $n\geq 1$, so that $A(w)=-c$, etc.

The payoff matrix for the three strategies *AllC*, *AllD*, and *TFT* is given by

$$\begin{pmatrix} b-c & -c & b-c \\ b & 0 & b(1-w) \\ b-c & -c(1-w) & b-c \end{pmatrix}, \tag{3.11}$$

where we omitted the factor $(1-w)^{-1}$, (i.e., considered the average payoff *per round*). Setting $w=1$ yields the infinitely repeated case.

In the general Prisoner's Dilemma game, the payoff matrix corresponding to matrix (3.11) is

$$\begin{pmatrix} R & S & R \\ T & P & (1-w)T + wP \\ R & (1-w)S + wP & R \end{pmatrix}. \tag{3.12}$$

3.4 PYRRHIC VICTORIES

Let us stick with the Donation game and normalize the corresponding replicator equation such that P_y, the payoff for defectors, is 0. Then we obtain

$$P_x = -c + wbz \qquad P_z = P_x + wcy. \tag{3.13}$$

We note that $P_z - \bar{P} = yg$, with

$$g = w(b-c)z - c(1-w). \tag{3.14}$$

On the edge of the state space simplex S_3 with $z = 0$ (no reciprocators), *AllD* clearly dominates. On the edge with $x = 0$, i.e., in a population consisting of defectors and TFT players, we have bi-stable dynamics. The unstable equilibrium is $F_{yz} = (0, 1 - \hat{z}, \hat{z})$, with

$$\hat{z} = \frac{(1-w)c}{w(b-c)}, \tag{3.15}$$

provided $\hat{z} < 1$, i.e., $w > c/b$. In particular, *TFT* is risk-dominant (see section 2.10) if

$$w > \frac{2c}{b+c}, \tag{3.16}$$

and selectively advantageous (see section 2.17) if

$$w > \frac{3c}{b+2c}. \tag{3.17}$$

Since $\hat{z}$ is small if w is close to 1, a small *TFT* population is able to invade a population of defectors if w, i.e., the "shadow of the future" is sufficiently large.

The edge $y = 0$ consists of fixed points only. Clearly, a population of *AllC* and *TFT* players will always cooperate, and none of the two strategies is favored. On the edge $y = 0$, those points with $z \geq c/wb$ are Nash equilibria, and the others are not. To see this, we have only to look at the sign of $P_y - \bar{P}$, i.e., of $P_x = -c + wbz$, and recall from section 2.10 that the Nash equilibria are exactly those fixed points that are saturated (i.e., if $y = 0$, then $P_y \leq \bar{P}$).

The other Nash equilibria of the game are the vertex $y = 1$ (defectors only) and the point F_{yz}. In the interior of the simplex S_3, there is no fixed point, since $P_z > P_x$ whenever $y > 0$. It is easy to see that the function

$$V = x^{\frac{1-w}{w}} z^{-\frac{1}{w}} g \tag{3.18}$$

is an invariant of motion, i.e., satisfies $\dot{V} = 0$.

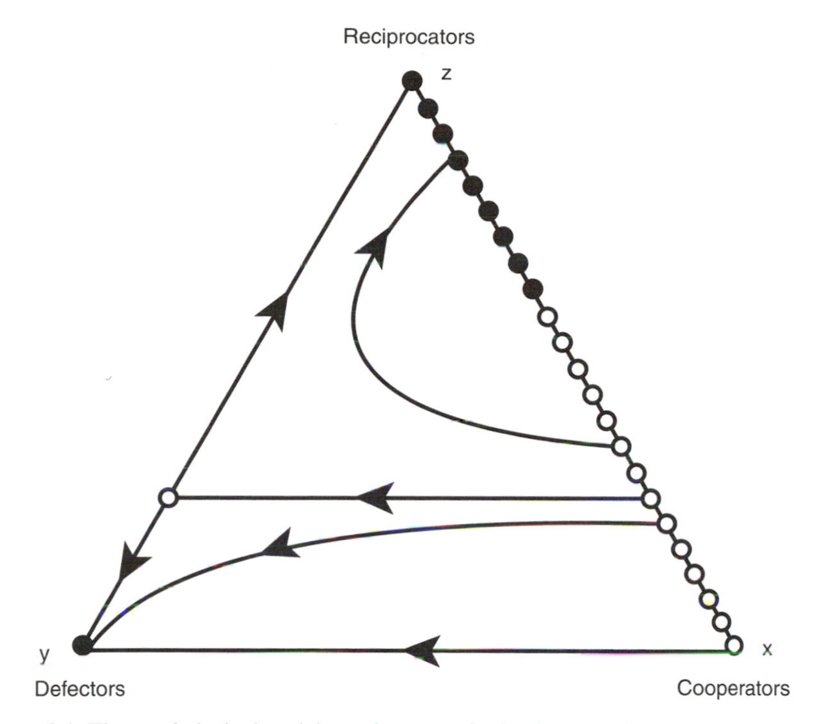

Figure 3.1 The good, the bad, and the reciprocator, in the absence of errors. A horizontal line $z = c/wb$ divides the state space. Below the line, defectors win; above the line, defectors are eliminated. Here and in all other figures, filled circles correspond to stable rest points, and empty circles to unstable rest points.

In the case $c < wb$, the dynamics shows an interesting behavior, see figure 3.1. The segment with $g = 0$ consists of a single orbit parallel to the edge $z = 0$, which converges to the saddle point F_{yz} and separates the simplex into two parts. Below this line, z decreases, and y converges to 1, i.e., defectors win. Above the line $g = 0$, z increases, and y converges to 0, i.e., defectors lose.

In the absence of defectors, any mixture of *TFT* players and *AllC* players corresponds to a rest point. Such a mixture can be viewed as a mixture of discriminating and indiscriminating altruists. If we assume that occasionally, small random shocks perturb the system, then these will send the system up and down the defectors-free edge $y = 0$. If a random shock introduces a small amount of defectors while $z > c/wb$, the defectors will forthwith be eliminated. If the defectors are introduced while $z < (1 - w)c/w(b - c)$, they will take over. But if the defectors are introduced in the "middle zone" where

$$c/wb > z > (1 - w)c/w(b - c), \tag{3.19}$$

the amount of defectors will first increase, and then vanish. During the phase of their invasion, the *AllD* players will exploit and eventually deplete the *AllC* players. This is a kind of Pyrrhic victory: the defectors end up meeting mostly *TFT* players, and this is their undoing.

Looking at it from the point of view of defectors, any invasion attempt while $z > \hat{z}$ is doomed to failure and will result in a state with $y = 0$ and $z > c/wb$. Figuratively speaking, the only hope for the defectors is to wait with their invasion attempt until drift, i.e., a succession of small random shocks, has moved the population state along the edge $y = 0$, to the region where $z < \hat{z}$. This drift needs time. If the invasion attempts occur too often, the drift will never have enough time to lead into the zone that favors the defectors. Thus defectors should not try to invade too often. In other terms, cooperators are safe only if invasion attempts by defectors are sufficiently frequent. If the invasion attempts are too rare, a cooperative society can lose its immunity—random fluctuations can lead to a population state with too few reciprocators to repel an invading minority of defectors.

3.5 REACTIVE STRATEGIES

So far, we have assumed that the players execute their intentions faultlessly. If we assume that they occasionally commit errors, we obtain very different results. This leads to the investigation of stochastic strategies, described by the probabilities, in each round, to cooperate or not.

To begin with, let us consider strategies given by triplets (f, p, q), where f is the probability to cooperate in round 0, and p resp. q are the probabilities to cooperate after a cooperation resp. defection by the co-player in the previous round. For such *reactive strategies*, the propensity to cooperate depends uniquely on what the co-player did in the previous round. The pair (p, q) defines the *reaction norm* of the strategy. It is a point in the unit square $[0, 1]^2$, and it is said to be *deterministic* if it corresponds to one of the corners. For instance, *TFT* corresponds to $(1, 1, 0)$ and *AllD* to $(0, 0, 0)$; both have deterministic reaction norms. A (small) probability ϵ to implement the unintended move would change this to $(1-\epsilon, 1-\epsilon, \epsilon)$ resp. $(\epsilon, \epsilon, \epsilon)$. We shall use the notation $\rho := p - q$. Clearly $| \rho | < 1$ except for some strategies with deterministic reaction norm, such as *TFT*.

Let us consider an (f, p, q) player encountering an (f', p', q') player. In each round, there are four possible outcomes, namely (C, C), (C, D), (D, C), and (D, D), depending on the moves of the first and the second player. This outcome can also be described by the payoff obtained by the first player, namely R, S, T, or P, which we enumerate by 1, 2, 3, 4. (Note that an S for the first player corresponds to a T for the second player.)

In the initial round, the probabilities $x_i(0)$ for outcome $i \in \{1, 2, 3, 4\}$ are given by the quadruple

$$\mathbf{x}(0) = (ff', f(1-f'), (1-f)f', (1-f)(1-f')). \tag{3.20}$$

In the following rounds, these probabilities change according to the reaction norms of the two players. We denote by p_{ij} the probability that from one round to the next, the state changes from i to j (with $i, j \in \{1, 2, 3, 4\}$). Thus $\mathbf{x}(n)$ turns into $\mathbf{x}(n+1)$ according to the transition rule

$$\mathbf{x}(n+1) = \mathbf{x}(n)\mathbf{P}, \tag{3.21}$$

where $\mathbf{P} = (p_{ij})$ is the stochastic matrix

$$\mathbf{P} = \begin{pmatrix} pp' & p(1-p') & (1-p)p' & (1-p)(1-p') \\ qp' & q(1-p') & (1-q)p' & (1-q)(1-p') \\ pq' & p(1-q') & (1-p)q' & (1-p)(1-q') \\ qq' & q(1-q') & (1-q)q' & (1-q)(1-q') \end{pmatrix}. \tag{3.22}$$

This yields a Markov chain.

3.6 LINKAGE

If the probabilities x_i to be in state i satisfy the condition

$$x_1 x_4 = x_2 x_3, \tag{3.23}$$

then the moves of the two players are independent. Indeed, x_1 is the probability that both player I *and* player II play C. The probability that I plays C is $x_1 + x_2$, and the probability that II plays C is $x_1 + x_3$. Independence means that $x_1 = (x_1 + x_2)(x_1 + x_3)$, which for $\mathbf{x} \in S_4$ is equivalent with $x_1 x_4 = x_2 x_3$. In this sense the *linkage* $D = x_1 x_4 - x_2 x_3$ measures the interdependence of the two players: $D = 0$ means that their moves are independent.

A straightforward computation shows that

$$D(n+1) = \rho\rho' D(n), \tag{3.24}$$

where $\rho = p - q$ and $\rho' = p' - q'$ as before. Indeed, we have only to replace $x_j(n+1)$ in $D(n+1)$ by $\sum_i x_i(n) p_{ij}$, using equation (3.21), and then compare the coefficients of the product terms $x_i(n) x_j(n)$. Most of the coefficients cancel obligingly, since $p_{k1} p_{k4} = p_{k2} p_{k3}$ and $p_{1k} p_{4k} = p_{2k} p_{3k}$ for $k = 1, 2, 3, 4$.

It follows that the *linkage disequilibrium* $D(n)$, which is 0 in the initial round, remains 0. (If it were initially distinct from 0, it would converge to 0 exponentially if at least one of the reaction norms is non-deterministic.) This confirms that the moves of the two players are independent in every round, as expected.

3.7 COOPERATION LEVELS

Players using reactive strategies play a kind of ping-pong with each other: if player II cooperates with a probability y, then player I cooperates with probability

$$\alpha(y) = py + q(1-y) = q + \rho y \tag{3.25}$$

in the following round. Thus if player I's *cooperation level* in round n is denoted by $c_n = x_1(n) + x_2(n)$, then

$$c_{n+2} = q + \rho(q' + \rho' c_n) = A + u c_n, \tag{3.26}$$

where $u := \rho\rho'$ and $A = q + \rho q'$ (which is $\alpha\alpha'(0)$). Equation (3.26) defines an affine-linear mapping from the unit interval (the set of all cooperation levels) into itself. The mapping can be iterated, starting from the initial round:

$$c_0 = f \mapsto c_2 = A + uf \mapsto c_4 = A + u(A + uf) \\ = A(1+u) + u^2 f \mapsto \cdots . \tag{3.27}$$

Since

$$c_{2n} = A(1 + u + \cdots + u^{n-1}) + u^n f = \frac{A}{1-u} + u^n \left(f - \frac{A}{1-u}\right), \tag{3.28}$$

we obtain

$$c_{2n} = v + u^n(f - v), \tag{3.29}$$

where

$$v := \frac{A}{1-u} = \frac{q + \rho q'}{1 - \rho\rho'} \tag{3.30}$$

is just the fixed point of $y \mapsto A + uy$. A similar equation holds for c_{2n+1} (with f replaced by $c_1 = q + \rho f'$). The cooperation level c_n thus converges to v. The same holds for the other player, whose cooperation level converges to v'. Clearly, one has $\alpha(v') = v$, etc. It is only if both strategies have deterministic reaction norms that the cooperation levels may periodically oscillate forever, for instance if a *TFT* player encounters a "suspicious *TFT* player" using $(0, 1, 0)$.

In addition to the stationary cooperation levels v and v' of the two players against each other, we can also consider the hypothetical cooperation levels s and s' which the players would obtain, in the limit, against a co-player using their own strategy. An (f, p, q) player reaches a cooperation level

$$s := \frac{q}{1-\rho} \tag{3.31}$$

against another (f, p, q) player. Interestingly, $v - v'$ has the same sign as $s - s'$ (and as $v - s'$, as well as $s - v'$). In particular, if two of the limits v, v', s, and s' of cooperation levels are equal, so are all four. It is useful to note that

$$v - s = \rho(v' - s). \tag{3.32}$$

This leads to a simple interpretation linking cooperation levels to reaction norms. All reaction norms (p', q') lying on the line from (p, q) to $(1, 0)$ (the *TFT* norm) have the same asymptotic cooperation level against themselves, and consequently against each other. If a reaction norm (p', q') lies above the line from (p, q) to $(1, 0)$, it has a higher asymptotic cooperation level (against itself, and against (p, q)), and vice versa.

3.8 PAYOFF VALUES

We shall not consider the case $u^2 = 1$, (which occurs only if both strategies have a deterministic reaction norm).

Since the decisions of the two players are independent, the player using (f, p, q) obtains in round n against a player using (f', p', q') the payoff

$$A(n) = Rc_n c'_n + Sc_n(1 - c'_n) + T(1 - c_n)c'_n + P(1 - c_n)(1 - c'_n). \quad (3.33)$$

In the special case of the Donation game, this reduces to

$$A(n) = bc'_n - cc_n. \quad (3.34)$$

For the infinitely iterated case $w = 1$ this means that the average payoff per round is

$$(R - S - T + P)vv' + (S - P)v + (T - P)v' + P, \quad (3.35)$$

which reduces to

$$bv' - cv \quad (3.36)$$

for the Donation game. These expressions do not depend on the initial propensities to cooperate, namely f and f'.

In order to obtain the total payoff for the Donation game with $w < 1$, we have to compute $\sum w^n c_n$. By equation (3.29),

$$\sum w^n c_n = \sum w^{2n}[v + u^n(f - v)] + \sum w^{2n+1}[v + u^n(c_1 - v)], \quad (3.37)$$

which is, up to the factor $[(1 - w)(1 - uw^2)]^{-1}$,

$$\begin{aligned} & v(1 - uw^2) - v(1 - w) - vw(1 - w) + f(1 - w) + c_1 w(1 - w) \quad (3.38)\\ &= vw^2(1 - u) + (1 - w)(f + qw + w\rho f')\\ &= (q + \rho q')w^2 + (1 - w)(f + qw + w\rho f'). \quad (3.39) \end{aligned}$$

Collecting the terms in f, f', q, and q', and setting $e := (1 - w)f + wq$, $e' := (1 - w)f' + wq'$, we obtain

$$\sum w^n c_n = \frac{e + w\rho e'}{(1 - w)(1 - uw^2)}. \quad (3.40)$$

Thus the average payoff per round is given by

$$\frac{-c(e + w\rho e') + b(e' + w\rho' e)}{1 - uw^2}. \quad (3.41)$$

3.9 THE GOOD, THE BAD, AND THE RECIPROCATOR WITH ERRORS

We will assume that an intended donation can fail with probability ϵ, and an intended refusal with probability $k\epsilon$, for some $k \geq 0$. It makes sense to distinguish between these two errors in implementation, and in particular to keep the case $k=0$ in mind. For instance, players who want to donate, but are out of funds, are failing to implement their intention. But it is unlikely that players who do not want to give anything away are absentminded enough to donate. Thus the three strategies *AllC*, *AllD* and *TFT* are now given by $\mathbf{e}_1=(1-\epsilon, 1-\epsilon, 1-\epsilon)$, $\mathbf{e}_2=(k\epsilon, k\epsilon, k\epsilon)$ and $\mathbf{e}_3=(1-\epsilon, 1-\epsilon, k\epsilon)$, respectively.

Applying expression (3.41) to these three strategies, we obtain a 3×3 payoff matrix M which, at first glance, looks somewhat daunting. But it can be simplified considerably, especially as the ρ values of the two unconditional strategies are 0 (i.e., $p=q$). Once more we use the fact that the replicator dynamics in S_3 is unchanged if we subtract, in each column of M, the diagonal from all elements. Up to the multiplicative factor $c(1-(k+1)\epsilon)$, the normalized matrix of payoff values per round is of the form

$$M = \begin{pmatrix} 0 & -1 & \delta\sigma \\ 1 & 0 & -\kappa\sigma \\ \delta & -\kappa & 0 \end{pmatrix} \tag{3.42}$$

where we used

$$\delta := w\epsilon, \quad \kappa := 1 - w + wk\epsilon, \quad \sigma := \frac{b\theta - c}{c - c\theta}, \text{ and } \theta = w(1-(k+1)\epsilon). \tag{3.43}$$

We note that $\bar{P} = z(1+\sigma)P_z$. Using

$$P_z - \bar{P} = P_z[1-(1+\sigma)z], \tag{3.44}$$

we see that in the interior of S_3, $\dot{z}=0$ holds whenever $g := 1-(1+\sigma)z$ vanishes. It is easy to see that $g=0$ corresponds to an orbit connecting the fixed points $F_{yz} := (0, 1-\hat{z}, \hat{z})$ and $F_{xz} := (1-\hat{z}, 0, \hat{z})$, where $\hat{z} := (1+\sigma)^{-1}$. On the edge $x=0$ defectors and reciprocators are engaged in a bi-stable competition, their basins of attraction separated by F_{yz}. On the edge $y=0$, reciprocators and *AllC* players are stably coexisting at the point F_{xz}. On the edge $z=0$ of unconditional players, the defectors dominate the cooperators.

In the interior of S_3 we obtain an invariant of motion

$$V := x^A y^B z^C [1-(1+\sigma)z] \tag{3.45}$$

with $A=\kappa/\theta$, $B=\delta/\theta$, and $C=-1/\theta$ (note that $A+B+C+1=0$). The interior rest point is

$$F = \frac{1}{1+\sigma(\kappa+\delta)}(\kappa\sigma, \delta\sigma, 1). \tag{3.46}$$

The dynamics is shown in figure 3.2. There is a horizontal orbit on the line $z=\hat{z}$, connecting the fixed points F_{xz} and F_{yz} (the latter is a Nash equilibrium). Below this line, all orbits converge to $y=1$, the defectors win. The part above the line is filled with periodic orbits surrounding the unique fixed point: they correspond to

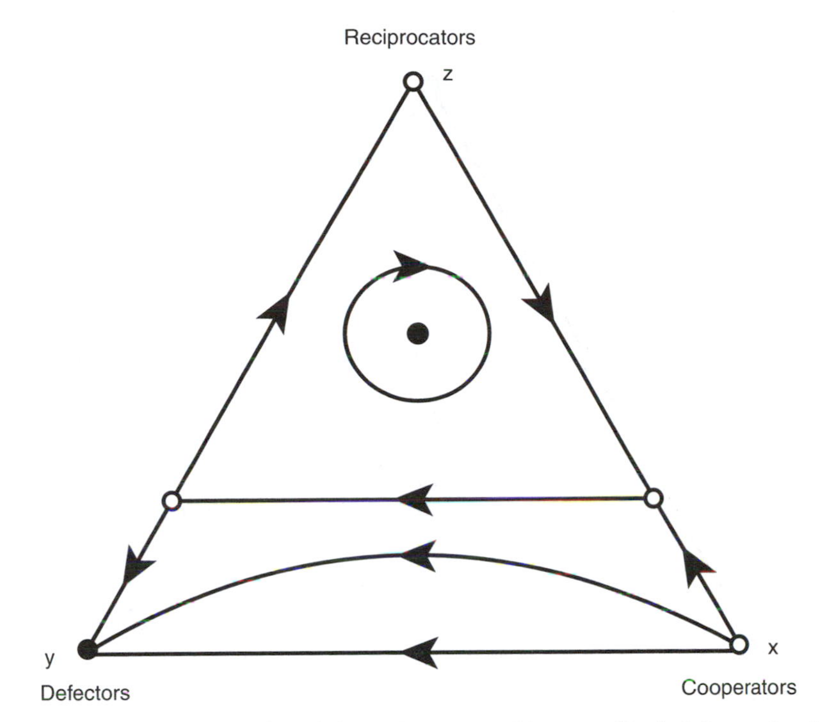

Figure 3.2 The good, the bad, and the reciprocator, with errors. If z is below a threshold, defectors win; if z is above the threshold, all three strategies co-exist, their frequencies oscillating periodically.

the constant level curves of the invariant of motion V given by expression (3.45). The time averages correspond to the values at the rest point F. This rest point is stable, but not asymptotically stable. We note that the amount of defectors at F can be made arbitrarily small if the error rate ϵ is sufficiently reduced. On the other hand, the basin of attraction of the *AllD* state ($y = 1$) can be arbitrarily small if w is sufficiently close to 1.

3.10 LIMITING CASES

For $w = 1$ we obtain, up to the multiplicative factor $c(1 - (k + 1)\epsilon)$, the payoff matrix

$$M = \begin{pmatrix} 0 & -1 & \beta \\ 1 & 0 & -k\beta \\ \epsilon & -k\epsilon & 0 \end{pmatrix} \tag{3.47}$$

where

$$\beta := \frac{1}{c}\left(\frac{b - c}{1 + k} - \epsilon b\right). \tag{3.48}$$

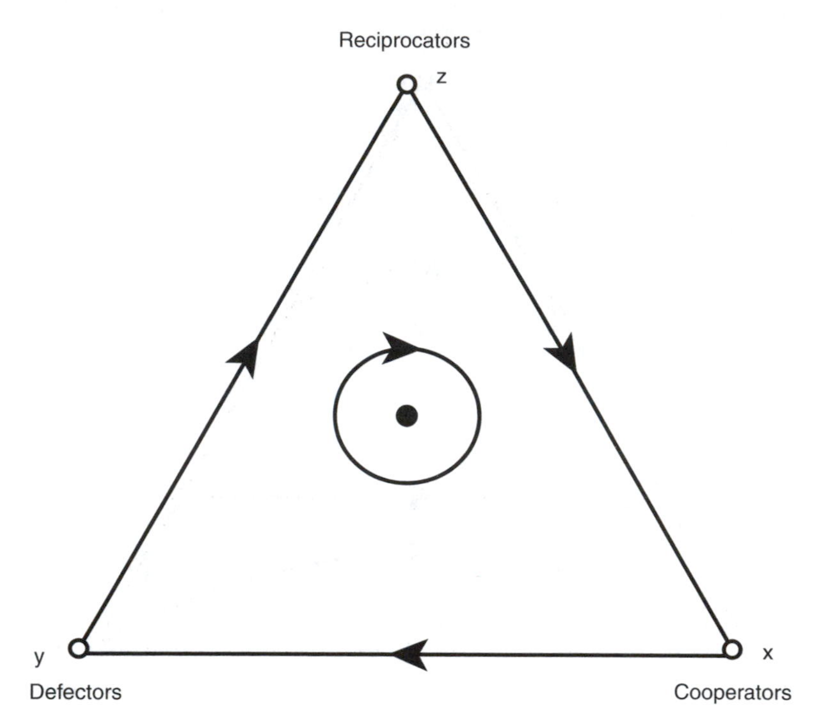

Figure 3.3 The infinitely iterated Donation game ($w = 1$), if there is a positive probability that intended moves (donation or refusal) are mis-implemented.

If $k > 0$ (i.e., if there is a positive probability for a donation, even if a refusal is intended), the dynamics is the same as in figure 3.2, the z coordinate of the separatrix is given by

$$\hat{z} := \frac{c}{(b-c)} \frac{(k+1)\epsilon}{(1-(k+1)\epsilon)}. \tag{3.49}$$

If $\epsilon \to 0$ the separatrix merges with the edge $z = 0$ and we obtain a system whose payoff matrix is

$$M = \begin{pmatrix} 0 & -c & (b-c)/(1+k) \\ c & 0 & -k(b-c)/(1+k) \\ 0 & 0 & 0 \end{pmatrix}. \tag{3.50}$$

This is a Rock-Scissors-Paper game: *AllD* is out-competed by *TFT*, which is out-competed by *AllC*, which is out-competed by *AllD* in turn. The unique rest point in the interior of S_3 is $F = \frac{1}{b}(k(b-c)/(k+1), (b-c)/(k+1), c)$. The replicator dynamics is as in figure 3.3.

If, on the other hand, we first consider the limiting case $\epsilon = 0$ (with $w < 1$), we obtain the dynamics shown in figure 3.1 If we then let w converge to 1, we obtain fig. 3.4. We note that the limits $w \to 1$ and $\epsilon \to 0$, therefore, do not commute.

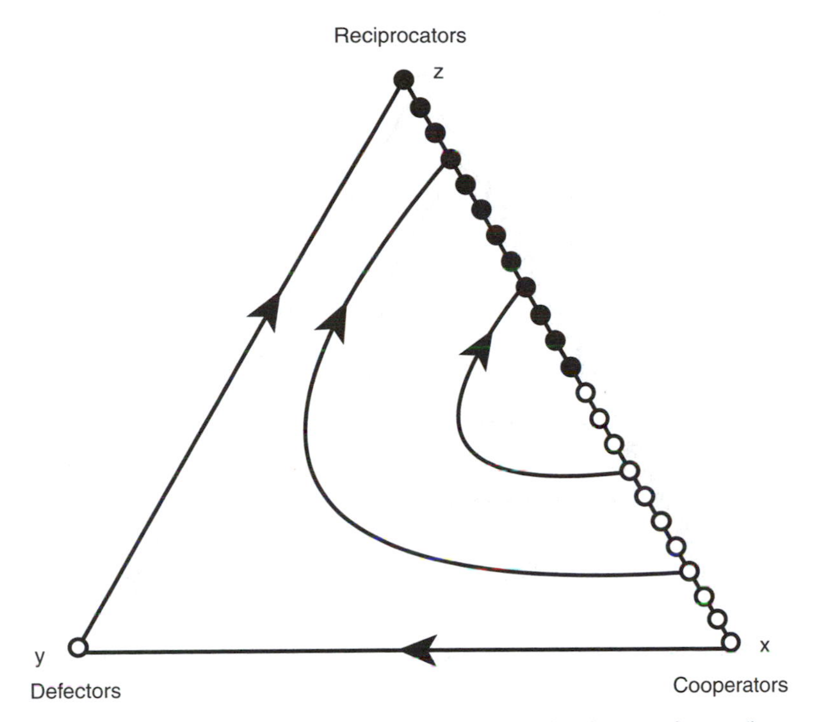

Figure 3.4 The infinitely iterated Donation game ($w = 1$), in the absence of errors (i.e., $\epsilon = 0$).

Suppose now that $k = 0$, i.e., that an intended refusal never fails. This is not without plausibility. In the limiting case $w = 1$, the payoff matrix is given, up to the factor $c(1 - \epsilon)$, by

$$M = \begin{pmatrix} 0 & -1 & \beta \\ 1 & 0 & 0 \\ \epsilon & 0 & 0 \end{pmatrix} \tag{3.51}$$

with $\beta = [(1 - \epsilon)b - c]/c$. This yields a completely different picture. The edge $x = 0$ consists of fixed points. Intuitively, this is clear: errors between two *TFT* players will eventually lead to mutual defection in each round, and this can never be redressed by another error. Thus the *TFT* players' average payoff per round will be 0. The rest points with $z \leq \bar{z}$ are Nash equilibria, with

$$\bar{z} = c/[b(1 - \epsilon)]. \tag{3.52}$$

The dynamics looks as in figure 3.5, which is an intriguing mirror-image of figure 3.1.

3.11 ADAPTIVE DYNAMICS

The reactive strategies (f, p, q) form a continuum. A heterogeneous population consisting of three or four such strategies can have a complex dynamics displaying

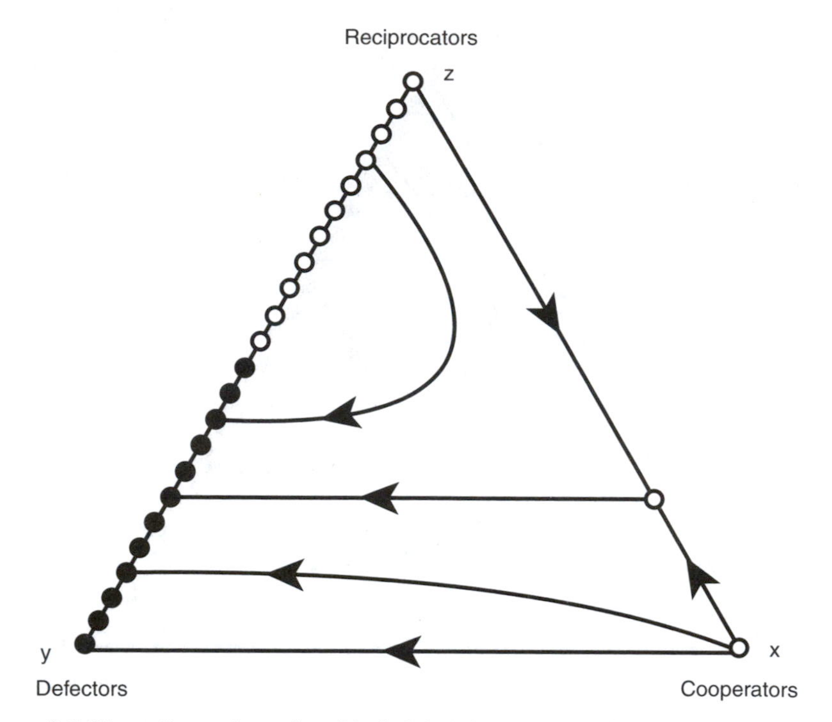

Figure 3.5 The replicator dynamics of the infinitely iterated Donation game, if only donations can be mis-implemented, but refusals are not. Cooperation vanishes in the long run.

limit cycles, heteroclinic cycles, or chaotic oscillations. Rather than pursue this point, let us ask how a *homogeneous* population evolves.

More precisely, we shall assume that the resident population is homogeneous, and that from time to time, a small minority of another type enters. These dissidents can do better or less well than the residents. Imitation will occur, and usually lead either to the elimination or to the fixation of this new type. After this, another minority can try its luck, etc. Such a limiting situation (with very rare innovations and strong imitation, or in a biological framework with very rare mutations and strong selection) can be described by a sequence of homogeneous populations. We shall describe an *adaptive dynamics* pointing towards the most favorable direction of evolution.

Let us first consider the limiting case $w = 1$. If we denote with $\mathbf{n} := (p, q)$ the reaction norm of the resident type and with $\mathbf{n}' := (p', q')$ that of the rare invading minority, we have to check whether invaders or the resident population are doing better. Individuals of both types are essentially interacting with the resident (since the dissidents are rare). Let $A(\mathbf{n}', \mathbf{n})$ be the average payoff of a player using the strategy $\mathbf{n}'$ against a player using $\mathbf{n}$. Hence the type $\mathbf{n}'$ can invade if and only if the payoff difference $A(\mathbf{n}', \mathbf{n}) - A(\mathbf{n}, \mathbf{n})$ is positive.

Let us denote, as in section 3.7, the asymptotic cooperation level of a (p, q) player against another (p, q) player by s, and the asymptotic levels of cooperation

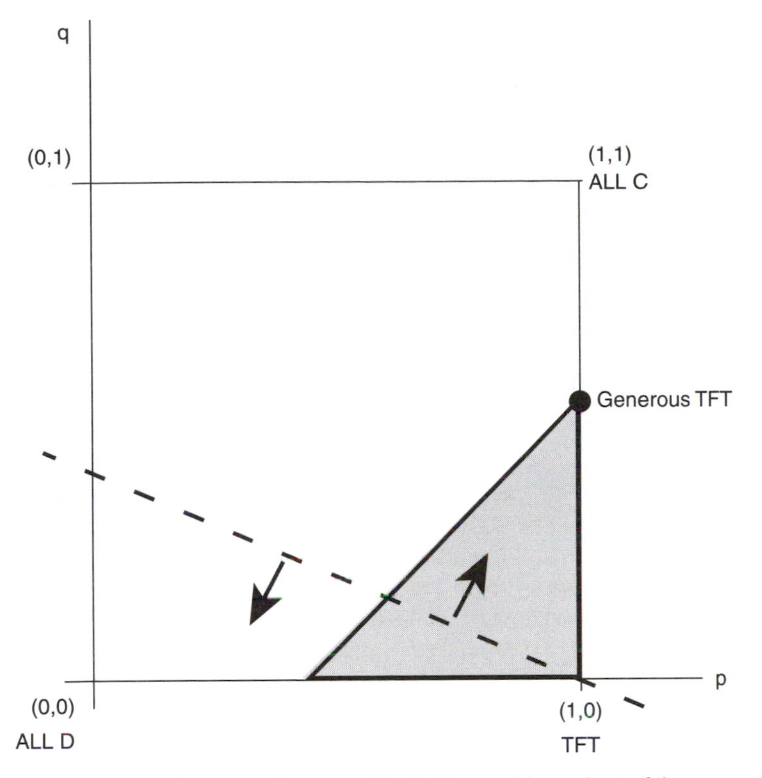

Figure 3.6 The cooperation-rewarding zone (shaded in grey) is a subset of the space of reaction norms (p, q) for the infinitely iterated Donation game. The arrows point in the direction of the most favorable adaptation. This direction is always orthogonal to the line connecting the norm with (1, 0).

between a (p, q) player and a (p', q') player by v resp. v'. From equation (3.34) we see that

$$A(\mathbf{n}, \mathbf{n}) = bs - cs \tag{3.53}$$

and

$$A(\mathbf{n}', \mathbf{n}) = bv - cv'. \tag{3.54}$$

Hence, using $\rho = p - q$, we obtain

$$A(\mathbf{n}', \mathbf{n}) - A(\mathbf{n}, \mathbf{n}) = (v' - s)(b\rho - c). \tag{3.55}$$

The line $\rho = c/b$, i.e., $q = p - (c/b)$, divides the square $[0, 1]^2$ of reaction norms (p, q) into two regions (see figure 3.6), namely the southeast corner, (which includes the *TFT* strategy (1, 0)) and the rest. As mentioned in section 3.7, the sign of $v' - s$ is positive resp. negative depending on whether $\mathbf{n}' = (p', q')$ lies above or below the line from $\mathbf{n} = (p, q)$ to $(1, 0)$. It follows that if $\mathbf{n}$ lies in the southeast corner, then precisely those strategies $\mathbf{n}'$ that are more cooperative can invade: indeed, if $s' > s$, then $v' > s$ and the invader's payoff is larger than that of the resident. We denote this region as the *cooperation-rewarding* zone. Conversely, if the homogeneous population adopts a strategy $\mathbf{n} = (p, q)$ that does not lie in this cooperation-rewarding zone, then every less cooperative strategy can invade. If $\mathbf{n}$ lies on the boundary of the cooperation-rewarding zone, i.e., satisfies $\rho = c/b$, then all strategies do exactly as well, against $\mathbf{n}$, as $\mathbf{n}$ does against itself.

If the invader's strategy $\mathbf{n}'$ is close to the resident's strategy $\mathbf{n}$, we can approximate the invader's payoff difference $A(\mathbf{n}', \mathbf{n}) - A(\mathbf{n}, \mathbf{n})$ by its first-order Taylor expansion, i.e., by

$$(p' - p)\frac{\partial A}{\partial p'}(\mathbf{n}', \mathbf{n}) + (q' - q)\frac{\partial A}{\partial q'}(\mathbf{n}', \mathbf{n}), \tag{3.56}$$

where the partial derivatives of the function $\mathbf{n}' \mapsto A(\mathbf{n}', \mathbf{n})$ are evaluated at $\mathbf{n}' = \mathbf{n}$. We accordingly define the adaptive dynamics in the space $[0, 1]^2$ of reaction norms (p, q) as

$$\dot{p} = \frac{\partial A}{\partial p'}(\mathbf{n}', \mathbf{n}) \qquad \dot{q} = \frac{\partial A}{\partial q'}(\mathbf{n}', \mathbf{n}), \tag{3.57}$$

where the derivatives are evaluated at $\mathbf{n}' = \mathbf{n}$. This yields a vector field pointing, for every homogeneous state $\mathbf{n}$, into the direction that is most advantageous for the invader. A straightforward computation yields the derivatives of $A(\mathbf{n}', \mathbf{n}) = bv - cv'$. One obtains

$$\dot{p} = q\frac{b\rho - c}{(1 - \rho)(1 - \rho^2)}, \tag{3.58}$$

$$\dot{q} = (1 - p)\frac{b\rho - c}{(1 - \rho)(1 - \rho^2)}. \tag{3.59}$$

Thus the vector $(\dot{p}, \dot{q})$ at the point $\mathbf{n} = (p, q)$ is orthogonal to the line from $\mathbf{n}$ to the *TFT* corner $(1, 0)$. In the cooperation-rewarding zone, and only there, this vector points upwards: if it pays to increase p (the gratitude), it pays to increase q (forgiveness), and vice versa.

The same holds for the general Prisoner's Dilemma case (if $w = 1$), except that the cooperation-rewarding zone is of a different shape: in equations (3.58) and (3.59), the term $b\rho - c$ is replaced by

$$(R - S - T + P)q\left(\frac{1 + \rho}{1 - \rho}\right) + (T - P)\rho + S - P. \tag{3.60}$$

There is no evolutionary tendency towards *TFT*: this strategy is a pivot, rather than a target, of adaptation.

3.12 GENEROUS TIT FOR TAT

Any strategy $\mathbf{n}$ at the boundary of the cooperation-rewarding zone, where $q = p - (c/b)$, has the property that every strategy $\mathbf{n}'$ yields the same payoff against $\mathbf{n}$, namely

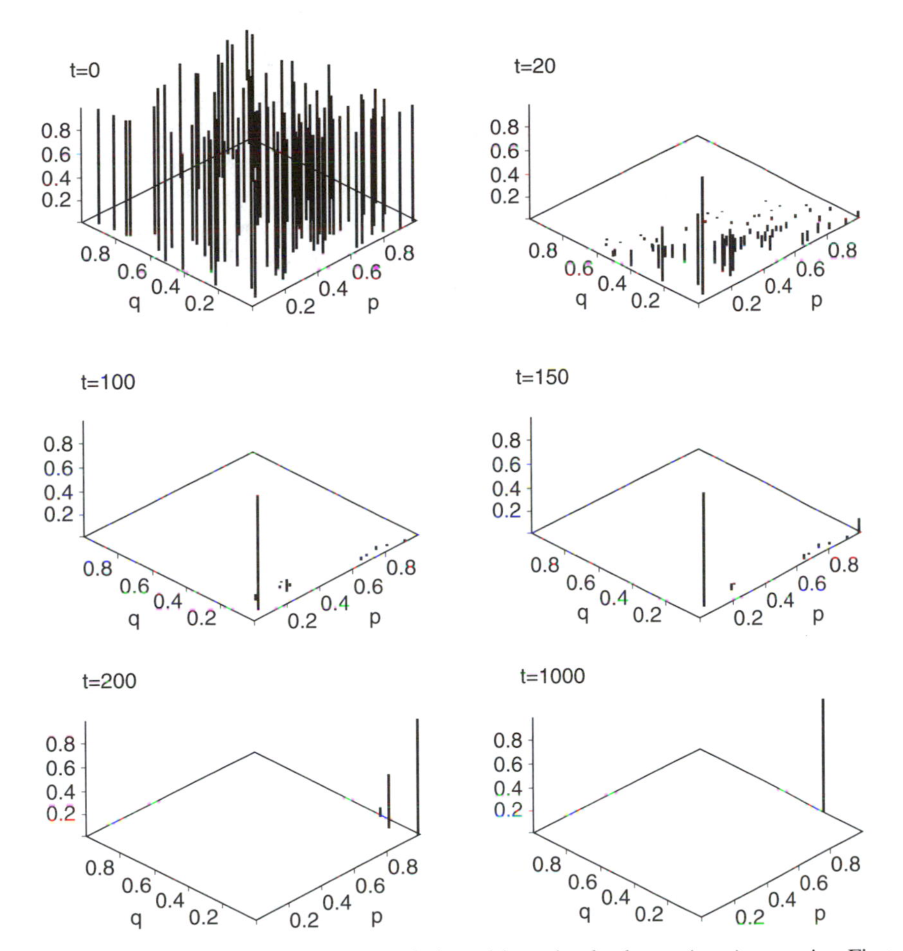

Figure 3.7 The evolution of a finite population with randomly chosen (p, q) strategies. First, "*AllD*" seems to win, then "*TFT*." But in the end, "*Generous TFT*" carries the day. (After Nowak and Sigmund (1992).)

bq. The largest such value, namely $b-c$, is obtained for $(1, 1-(c/b))$. This strategy is called *Generous TFT* (*GTFT*). A *Generous TFT* player always cooperates after a co-player's C, but does not always defect after a co-player's D. Rather, such a player forgives with a well-specified probability, namely $(b-c)/b$.

Generous TFT shows up in individual based computer simulations, see figure 3.7. Let us consider a large fictitious population of players who are assigned strategies chosen at random in the (p, q) square. Thus the initial population is not at all homogeneous. It can consist of hundreds of different types. Let us assume that players meet randomly and play a repeated Donation game against each other, with a large number of rounds. Let us furthermore assume that they update their strategy from time to time, by imitating more successful players. Quickly, most of the strategies will be eliminated from the population. In general, only three out of the initial set of strategies will play a role: those closest to *AllD*, *TFT*, and *GTFT*. We shall denote these approximations by "*AllD*", "*TFT*", and "*GTFT*", respectively. What one observes at first is a strong tendency towards "*AllD*." The other strategies seem hopelessly outclassed. But then, it frequently happens (for instance if "*TFT*" is below the line from "*AllD*" to *TFT*) that "*TFT*" experiences an upsurge, and displaces "*AllD*." But this is not the end of the story. The population has reached the cooperation-rewarding zone, and strategies that have higher p and q values can return. In particular, the more tolerant "*GTFT*" supersedes the stern "*TFT*," and becomes fixed in the population. The striking point is that "*GTFT*" on its own can never beat "*AllD*." It needs the catalytic action of "*TFT*." It seems almost like the succession of three social phases: first the "dog-eat-dog" world of *AllD*, then the "law of the talion" represented by *TFT* and finally the age of the tolerant, but not too tolerant *GTFT*.

Similar results hold for the adaptive dynamics of the Donation game if $w < 1$. In this case, the probability f to cooperate in the initial round is an additional trait. The adaptive dynamics at $\mathbf{n} = (f, p, q)$ is given by

$$\dot{f} = \frac{bw\rho - c}{1 - w^2\rho^2}. \tag{3.61}$$

$$\dot{p} = \dot{f}\left(\frac{w}{1-w}\right)\left(\frac{e}{1-w\rho}\right) \tag{3.62}$$

$$\dot{q} = \dot{f}\left(\frac{w}{1-w}\right)\left(1 - \frac{e}{1-w\rho}\right). \tag{3.63}$$

Here, $e = (1-w)f + wq$ as in section 3.8. Once again, all the components have the same sign (because $0 < e < 1 - w\rho$), so that we may again speak of a cooperation-rewarding zone. A direct computation shows that we can display the adaptive dynamics in a suggestive way:

$$\dot{f} = \frac{1-w}{1-w^2\rho^2}[A(AllC, \mathbf{n}) - A(AllD, \mathbf{n})] \tag{3.64}$$

$$\dot{p} = \frac{w}{1-w^2\rho^2}[A(\mathbf{n}, \mathbf{n}) - A(AllD, \mathbf{n})] \tag{3.65}$$

$$\dot{q} = \frac{w}{1-w^2\rho^2}[A(AllC, \mathbf{n}) - A(\mathbf{n}, \mathbf{n})]. \tag{3.66}$$

3.13 MEMORY-ONE STRATEGIES

So far, we have considered *reactive strategies* that depend only on the co-player's previous move. But it seems reasonable to assume that players also take their own move into account. It is probably easier to forgive a co-player's defection if it was matched by one's own defection, rather than if it exploited one's own cooperativeness. Hence we shall consider stochastic strategies $(f, q_R, q_S, q_T, q_P) \in [0, 1]^5$

where f, as before, is the propensity to play C in the initial round, and $q_R, q_S, \dots$ are the propensities to play C after having experienced an $R, S, \dots$ in the previous round.

Let us assume that player I using (f, q_R, q_S, q_T, q_P) encounters a co-player II using $(f', q'_R, q'_S, q'_T, q'_P)$. Again, we are dealing with a Markov chain; in every round, the state is specified by the payoff obtained by player I. The transition probabilities are given by the matrix

$$Q = \begin{pmatrix} q_R q'_R & q_R(1-q'_R) & (1-q_R)q'_R & (1-q_R)(1-q'_R) \\ q_S q'_T & q_S(1-q'_T) & (1-q_S)q'_T & (1-q_S)(1-q'_T) \\ q_T q'_S & q_T(1-q'_S) & (1-q_T)q'_S & (1-q_T)(1-q'_S) \\ q_P q'_P & q_P(1-q'_P) & (1-q_P)q'_P & (1-q_P)(1-q'_P) \end{pmatrix}, \quad (3.67)$$

(again one player's S is the other player's T).

The initial probabilities for the four states are given by the vector

$$\mathbf{x}(0) = (ff', f(1-f'), (1-f)f', (1-f)(1-f')), \quad (3.68)$$

which we denote by $\mathbf{f}$. In the next round, the probabilities are given by $\mathbf{f}Q$, and in round n by $\mathbf{f}Q^n$. For $n \geq 1$, the probabilities need no longer be in linkage equilibrium (the matrix Q satisfies $q_{k1}q_{k4} = q_{k2}q_{k3}$ but not $q_{1k}q_{4k} = q_{2k}q_{3k}$). If we denote by $\mathbf{g}$ the vector (R, S, T, P), then the payoff for player I in round n is given by

$$A(n) = \mathbf{g} \cdot \mathbf{f}Q^n. \quad (3.69)$$

For $w < 1$ the average payoff per round, as shown in equation (3.8), is $(1-w) \sum w^n A(n)$, i.e.,

$$(1-w)\mathbf{g} \cdot \mathbf{f}(Id - wQ)^{-1}, \quad (3.70)$$

where Id is the 4×4 identity matrix. For $w = 1$ we must proceed differently. If the matrix Q is mixing, i.e., if there exists an m such that all entries of Q^m are strictly positive, then there exists a unique vector $\pi \in S_4$ that is a left eigenvector of Q for the eigenvalue 1, i.e., $\pi = \pi Q$. The components π_R, π_S, π_T, and π_P denote the stationary probabilities of the four states, and we have

$$\mathbf{f}Q^n \to \pi \quad (3.71)$$

for every initial state $\mathbf{f}$. The average payoff per round, in this case, is $\mathbf{g} \cdot \pi$, which for the Donation game reduces to

$$b(\pi_R + \pi_T) - c(\pi_R + \pi_S). \quad (3.72)$$

3.14 THE SPACE OF REACTION NORMS

For $w = 1$ we can neglect the initial probability to cooperate and concentrate on the space of reaction norms (q_R, q_S, q_T, q_P). This unit cube is spanned by its sixteen corners, i.e., by the quadruples $(u_R, \dots, u_P)$ where u_i is 1 or 0 depending on whether the strategy prescribes to use C or D after outcome $i \in \{R, S, T, P\}$. We can number these strategies as S_j, where j ranges from 0 to 15 and is given, in binary notation, by $u_R u_S u_T u_P$. Hence $AllD = (0, 0, 0, 0)$ is S_0, $AllC = (1, 1, 1, 1)$ is S_{15}, $TFT = (1, 0, 1, 0)$ is S_{10}, etc. If we compute the transition matrix $\mathbf{P}$ for an S_i player meeting an S_j co-player, we find that in general it is not irreducible, let alone mixing: there are too many zeros, only one entry in each row does not vanish. Hence the stationary distributions are not uniquely determined.

This is different if we assume that every strategy is subject to errors in implementation: with a probability ϵ, the move is the opposite of what the strategy prescribes. Then each 1 turns into $1 - \epsilon$ and each 0 into ϵ. Strategy S_j turns into $S_j(\epsilon)$. For instance, *TFT*, i.e., $S_{10} = (1, 0, 1, 0)$, turns into $S_{10}(\epsilon) = (1 - \epsilon, \epsilon, 1 - \epsilon, \epsilon)$, etc.

It is straightforward to compute the payoff for strategy $S_i(\epsilon)$ against $S_j(\epsilon)$. The corresponding transition matrix is $Q(\epsilon)$, its elements are quadratic polynomials in ϵ. We can develop

$$Q(\epsilon) = Q + \epsilon Q_1 + \epsilon^2 Q_2, \tag{3.73}$$

where Q is a stochastic matrix with exactly one 1 in each row and Q_1 and Q_2 have row sums 0. We may view $Q(\epsilon)$ as a perturbation of the matrix Q and treat the problem of finding the left eigenvector $\mathbf{s}(\epsilon)$ of $Q(\epsilon)$ as a perturbation problem. Thus we set

$$\mathbf{s}(\epsilon) = \pi + \epsilon \mathbf{x} + \epsilon^2 \mathbf{y} + \cdots, \tag{3.74}$$

where the stochastic vector π is a solution of the unperturbed eigenvalue problem $\pi Q = \pi$, whereas the components of the vectors $\mathbf{x}$ and $\mathbf{y}$ must sum up to 0. By expanding $\mathbf{s}(\epsilon) Q(\epsilon) = \mathbf{s}(\epsilon)$ and comparing powers of ϵ, this yields not only the limiting value π for the payoff (if $\epsilon \to 0$), but also the first order term $\mathbf{x}$.

Let us consider, for example, $S_8 = (1, 0, 0, 0)$ against $S_{11} = (1, 0, 1, 1)$. S_8 is also called *Grim*, because it is a grim variant of *TFT*, prescribing to defect except after a round of mutual cooperation; whereas S_{11}, also know as *Firm But Fair* (*FBF*) is a tolerant brother of *TFT*, prescribing to play C if both players defected in the previous round. In that case,

$$Q = \begin{pmatrix} 1 & 0 & 0 & 0 \\ 0 & 0 & 1 & 0 \\ 0 & 0 & 0 & 1 \\ 0 & 0 & 1 & 0 \end{pmatrix}, \tag{3.75}$$

which is a reducible matrix, and

$$Q_1 = \begin{pmatrix} -2 & 1 & 1 & 0 \\ 1 & 0 & -2 & 1 \\ 0 & 1 & 1 & -2 \\ 1 & 0 & -2 & 1 \end{pmatrix}. \tag{3.76}$$

The equation $\pi Q = \pi$ yields $\pi_2 = 0$ and $\pi_3 = \pi_4$, i.e., $\pi = (1 - 2a, 0, a, a)$ for unknown a. The equation $\pi Q_1 + \mathbf{x} Q = \mathbf{x}$ yields $a = 2/5$, so that $\pi = (1/5, 0, 2/5, 2/5)$. We note that in this case, we did not need the ϵ^2 term, but sometimes we do.

In table 3.1 we display, for the Donation game, the resulting 16×16 matrix A, with a_{ij} denoting the payoff for an $S_i(\epsilon)$ player against a $S_j(\epsilon)$ player (or more

Table 3.1 The Simultaneous Donation Game with Errors in Implementation.

$*$	S_0	S_1	S_2	S_3	S_4	S_5	S_6	S_7	S_8	S_9	S_{10}	S_{11}	S_{12}	S_{13}	S_{14}	S_{15}
S_0	$\mathbf{0}$	$\frac{b}{2}$	0	$\frac{b}{2}$	$\frac{b}{3}$	b	$\frac{b}{2}$	b	0	$\frac{b}{2}$	0	$\frac{b}{2}$	$\frac{b}{2}$	b	$\frac{2b}{3}$	b
S_1	$-\frac{c}{2}$	$\mathbf{\frac{b-c}{2}}$	$\frac{b-c}{3}$	$\frac{b-c}{2}$	$\frac{b-2c}{5}$	$\frac{2b-c}{3}$	b	$\frac{3b-c}{4}$	$-\frac{c}{2}$	$\frac{2b-c}{3}$	$\frac{b-c}{3}$	$\frac{2b-c}{3}$	$\frac{2b-c}{4}$	b	b	b
S_2	0	$\frac{b-c}{3}$	$\mathbf{\frac{b-c}{4}}$	$\frac{b-c}{2}$	0	$\frac{2b-c}{3}$	0	$\frac{2b-c}{3}$	0	$\frac{b-c}{3}$	$\frac{b-c}{3}$	$\frac{b-c}{2}$	$\frac{2b-c}{4}$	$\frac{2b-c}{2}$	$\frac{4b-2c}{5}$	$\frac{2b-c}{2}$
S_3	$-\frac{c}{2}$	$\frac{b-c}{2}$	$\frac{b-c}{2}$	$\mathbf{\frac{b-c}{2}}$	$-\frac{c}{2}$	$\frac{b-c}{2}$	$\frac{b-c}{2}$	$\frac{b-c}{2}$	$-\frac{c}{2}$	$\frac{b-c}{2}$	$\frac{b-c}{2}$	$\frac{b-c}{2}$	$\frac{b-c}{2}$	$\frac{2b-c}{2}$	$\frac{2b-c}{2}$	$\frac{2b-c}{2}$
S_4	$-\frac{c}{3}$	$\frac{2b-c}{5}$	0	$\frac{b}{2}$	$\mathbf{\frac{b-c}{4}}$	b	$\frac{b}{3}$	b	$-\frac{c}{3}$	$\frac{2b-c}{5}$	0	$\frac{b}{2}$	$\frac{3b-c}{6}$	b	$\frac{2b}{3}$	b
S_5	$-c$	$\frac{b-2c}{3}$	$\frac{b-2c}{3}$	$\frac{b-c}{2}$	$-c$	$\mathbf{\frac{b-c}{2}}$	$\frac{b-c}{2}$	$\frac{2b-c}{3}$	$-c$	$\frac{b-c}{2}$	$\frac{b-c}{2}$	$\frac{2b-c}{3}$	$\frac{b-c}{2}$	b	b	b
S_6	$-\frac{c}{2}$	$-c$	0	$\frac{b-c}{2}$	$-\frac{c}{3}$	$\frac{b-c}{2}$	$\mathbf{0}$	$\frac{2b-c}{3}$	$-\frac{2c}{3}$	$-c$	$\frac{b-c}{2}$	$\frac{2b-2c}{3}$	$\frac{b-c}{2}$	$\frac{4b-3c}{5}$	$\frac{4b-2c}{5}$	$\frac{2b-c}{2}$
S_7	$-c$	$\frac{b-3c}{4}$	$\frac{b-2c}{3}$	$\frac{b-c}{2}$	$-c$	$\frac{b-2c}{3}$	$\frac{b-2c}{3}$	$\mathbf{\frac{b-c}{2}}$	$-c$	$-c$	$\frac{2b-2c}{3}$	$\frac{2b-2c}{3}$	$\frac{2b-3c}{4}$	$\frac{4b-3c}{5}$	$\frac{2b-c}{2}$	$\frac{2b-c}{2}$
S_8	0	$\frac{b}{2}$	0	$\frac{b}{2}$	$\frac{b}{3}$	b	$\frac{2b}{3}$	b	$\mathbf{0}$	$\frac{3b-c}{5}$	0	$\frac{3b-c}{5}$	$\frac{3b-c}{6}$	$\frac{3b-c}{3}$	$\frac{3b-c}{4}$	$\frac{3b-c}{3}$
S_9	$-\frac{c}{2}$	$\frac{b-2c}{3}$	$\frac{b-c}{3}$	$\frac{b-c}{2}$	$\frac{b-2c}{5}$	$\frac{b-c}{2}$	b	b	$\frac{b-3c}{5}$	$\mathbf{b-c}$	$\frac{b-c}{2}$	$b-c$	$\frac{b-c}{2}$	$\frac{3b-2c}{3}$	$\frac{3b-c}{3}$	$\frac{2b-c}{2}$
S_{10}	0	$\frac{b-c}{3}$	$\frac{b-c}{3}$	$\frac{b-c}{2}$	0	$\frac{b-c}{2}$	$\frac{b-c}{2}$	$\frac{2b-2c}{3}$	0	$\frac{b-c}{2}$	$\mathbf{\frac{b-c}{2}}$	$\frac{2b-2c}{3}$	$\frac{b-c}{2}$	$b-c$	$b-c$	$b-c$
S_{11}	$-\frac{c}{2}$	$\frac{b-2c}{3}$	$\frac{b-c}{2}$	$\frac{b-c}{2}$	$-\frac{c}{2}$	$\frac{b-2c}{3}$	$\frac{2b-2c}{3}$	$\frac{2b-2c}{3}$	$\frac{b-3c}{5}$	$b-c$	$\frac{2b-2c}{3}$	$\mathbf{\frac{3b-3c}{4}}$	$\frac{2b-3c}{4}$	$b-c$	$b-c$	$b-c$
S_{12}	$-\frac{c}{2}$	$\frac{b-2c}{4}$	$\frac{b-2c}{4}$	$\frac{b-c}{2}$	$\frac{b-3c}{6}$	$\frac{b-c}{2}$	$\frac{b-c}{2}$	$\frac{3b-2c}{4}$	$\frac{b-3c}{6}$	$\frac{b-c}{2}$	$\frac{b-c}{2}$	$\frac{3b-2c}{4}$	$\mathbf{\frac{b-c}{2}}$	$\frac{5b-3c}{6}$	$\frac{5b-3c}{6}$	$\frac{2b-c}{2}$
S_{13}	$-c$	$-c$	$\frac{b-2c}{2}$	$\frac{b-2c}{2}$	$-c$	$-c$	$\frac{3b-4c}{5}$	$\frac{3b-4c}{5}$	$\frac{b-3c}{3}$	$\frac{2b-3c}{3}$	$b-c$	$b-c$	$\frac{3b-5c}{6}$	$\mathbf{\frac{3b-3c}{4}}$	$\frac{3b-2c}{3}$	$\frac{3b-2c}{3}$
S_{14}	$\frac{-2c}{3}$	$-c$	$\frac{2b-4c}{5}$	$\frac{b-2c}{2}$	$\frac{-2c}{3}$	$-c$	$\frac{2b-4c}{5}$	$\frac{b-2c}{2}$	$\frac{b-3c}{4}$	$\frac{b-3c}{3}$	$b-c$	$b-c$	$\frac{3b-5c}{6}$	$\frac{2b-3c}{3}$	$\mathbf{b-c}$	$b-c$
S_{15}	$-c$	$-c$	$\frac{b-2c}{2}$	$\frac{b-2c}{2}$	$-c$	$-c$	$\frac{b-2c}{2}$	$\frac{b-2c}{2}$	$\frac{b-3c}{3}$	$\frac{b-2c}{2}$	$b-c$	$b-c$	$\frac{b-2c}{2}$	$\frac{2b-3c}{3}$	$b-c$	$\mathbf{b-c}$

precisely, its limit for $\epsilon \to 0$). We note an obvious symmetry: if $a_{ij} = xb - yc$, then $a_{ji} = yb - xc$.

If the resident population is playing S_0, i.e., *AllD*, then no strategy can invade except $S_2 = (0, 0, 1, 0)$, the "*Grim*" strategy $S_8 = (1, 0, 0, 0)$, and the *TFT* strategy $S_{10} = (1, 0, 1, 0)$. Since S_2 is dominated by S_{10} (in the absence of other strategies), this means that *TFT* can overcome *AllD*. But *TFT* can be superseded by more tolerant strategies, such as S_{15}, i.e., *AllC*, and these can in turn be displaced by *AllD*. However, this tendency to cycle can be broken up by S_9. This strategy dominates S_{10}, S_2, and (if $b > 3c$) also S_8, and it cannot be invaded by *AllD* as long as $b > 2c$, i.e., the cost-to-benefit ratio is less than $1/2$.

We note that S_9 is the only strategy that cannot be invaded by any other S_i (for $b > 3c$). Moreover, S_9 is very good against itself: a population of S_9 players earns $b - c$, which is the best a homogeneous population can achieve. Only S_{14} and S_{15} do as well, but these are easy prey to S_1 or S_0.

3.15 WIN-STAY, LOSE-SHIFT

The strategy $S_9 = (1, 0, 0, 1)$, for reasons difficult to fathom, is called *Pavlov*. It has the remarkable property of being *error-correcting*. If two players using *Pavlov* play against each other, they will cooperate most of the time. If player II, say, defects by mistake, then in the next round both players will play D, and thereafter resume mutual cooperation, like an old couple after a row (see fig. 3.8). Moreover, if a *Pavlov* player plays against *AllC*, it will shamelessly exploit the co-player. After the first accidental D, it will continue playing D until a further error occurs. This is an important property for safeguarding the population against eventual invasions by defectors. A *TFT* population, for instance, will quickly be subverted by *AllC* players, and these will be open to exploitation by *AllD*.

(a) Pavlov-player I   C C C ... C C D C C ...
    Pavlov-player II   C C C ... C D D C C ...
                                   ↗

(b) TFT-player I   C C C ... C C D C D ...
    TFT-player II   C C C ... C D C D C ...
                                   ↗

(c) ALL C-player I   C C C ... C C C C ...
    Pavlov-player II   C C C ... C D D D ...
                                   ↗

(d) ALL C-player I   C C C ... C C C C ...
    TFT-player II   C C C ... C D C C ...
                                   ↗

Figure 3.8 The effect of an erroneous defection in the iterated Prisoner's Dilemma game. The arrow denotes the mis-implemented move in each run.

Pavlov-player I	C	C	D	D
Player II	C	D	C	D
	↓	↓	↓	↓
Payoff for I	R	S	T	P
Next move for I	C	D	D	C

Figure 3.9 *Pavlov* as a Win-Stay, Lose-Shift strategy. After obtaining the larger payoff values T and R, a *Pavlov* player repeats the former, successful move. After obtaining the smaller payoff values P and S, the *Pavlov* player switches to the other move.

The strategy S_9 prescribes playing C if and only if, in the previous round, the co-player did the same as the other player. There is a suggestive property behind this mechanism, see figure 3.9. The strategy effectively repeats the previous move if it obtained a positive payoff (a reward, such as $b-c$, or better still the temptation b). It switches to the other move if the payoff was non-positive (payoff 0 if both players defected, or the sucker's payoff $-c$). This is the simplest conceivable learning mechanism, well-known to animal trainers and parents alike. *Win-Stay, Lose-Shift* is a wide-spread maxim of animal behavior.

The condition $b > 2c$ implies that *Pavlov* is not dominated by *AllD*, but that the two strategies are engaged in a bi-stable competition. The condition $b > 3c$ implies that *Pavlov* is risk-dominant.

It is interesting to consider finite populations in this context. Let us consider the two cases (a) $b = 5c/2$ and (b) $b = 4c$, a population size $M = 100$ and selection strength $s = 1/10$. Let us also assume the adiabatic case (very small innovation rates μ, see section 2.17). A population consisting only of the types $S_0 = AllD$ and $S_{10} = TFT$ will be dominated by *TFT*. (In the numerical example, *TFT* occurs with 97 percent in the stationary distribution given by expression (2.90) in case (a), and with 99 percent in case (b).) This reflects the fact that *TFT* dominates *AllD*. But if $AllC = S_{15}$ is also allowed in the population, then the stationary distribution is dominated by *AllD* (64 percent in case (a) and 66 percent in case (b)). Now let us consider a population with the strategies $S_9 = Pavlov$, *AllC*, and *AllD*. If $b > 3c$, *Pavlov* risk-dominates *AllD*. This corresponds to example (b), and we see indeed that the stationary distribution consists of 90 percent *Pavlov*. If $2c < b < 3c$, *AllD* risk-dominates *Pavlov*, and we find 80 percent of defectors in the stationary distribution. This changes dramatically if we also include *TFT*: in that case, example (a) leads to 50 percent *Pavlov* (and example (b) to 95 percent). Thus *TFT* is not the winner, but can act as a king-maker—decisive for the outcome of the contest between *AllD* and *Pavlov*.

In the general Prisoner's Dilemma game, *Pavlov* acts according to a threshold separating the two better outcomes T and R from the two worse outcomes P and S. A move yielding an outcome above the *aspiration level* is repeated, a move yielding an outcome below the aspiration level is not. One can consider other aspiration levels. If the aspiration level is more ambitious, content only with T, this leads to the strategy $S_1 = (0, 0, 0, 1)$, a bully-like strategy that only cooperates after a mutual defection. It relentlessly defects whenever it can exploit a sucker, but switches as soon as it meets a defection. This is an overly ambitious Win-Stay, Lose-Shift strategy, and it fails. Similarly, a more modest aspiration level (between P and S) leads to $S_8 = (1, 0, 0, 0)$, which is doing rather well, especially in a population of defectors. Finally, one could also view $S_3 = (0, 0, 1, 1)$ and $S_{12} = (1, 1, 0, 0)$ as extreme forms of Win-Stay, Lose-Shift strategies. The former switches its move from one round to the next, never satisfied by any outcome. The latter never switches except by mistake, and always repeats itself, apparently content with every outcome. Obviously, it is important to have the "right" aspiration level. *Pavlov* is, in this sense, the most balanced of all Win-Stay, Lose-Shift rules. It is also doing well in the iterated Snowdrift game, which is described in matrix (3.4).

3.16 AUTOMATA

Memory-one strategies with deterministic reaction norms can easily be implemented by finite state automata. For instance, *Pavlov* can be implemented by an automaton with only two inner states, which we shall denote by *Same* and *Diff*. The automaton is in state *Same* if in the previous round, the player experienced payoff R or P (i.e., both players cooperated, or both defected), and it is in state *Diff* otherwise. In each state, the automaton prescribes the next move, i.e., C or D. The analysis of *Pavlov* can then be performed very easily by means of a directed graph, see figure 3.10. The nodes of the graph are the states *Same* and *Diff* of the player. Two directed arrows are leaving from each node, one solid and the other dashed. The solid arrow describes the transition if the player uses the move prescribed by *Pavlov* (i.e., C in *Same* and D in *Diff*). The dashed arrow describes the transition if the other move is used. In both cases, the co-player is assumed to follow the *Pavlov* strategy. Along each arrow, one can see the corresponding payoff of the player. Clearly, it is best always to follow the solid arrow, if $2R > T + P$ (or, in the case $w < 1$, if $R + wR > T + wP$). For the Donation game, this reduces to the familiar condition $b > 2c$ (resp. $w(b - c) > c$). If this condition holds, it is always best, against a *Pavlov* player, to do what the *Pavlov* rule prescribes. In a population of *Pavlov* players, it is best to follow suit. Similar graphs can be studied for all memory-one strategies. In general, this will be more complicated than for *Pavlov*, where the two players are always in the same state. But the four states (C, C), (C, D), (D, C) and (D, D) will always be enough to

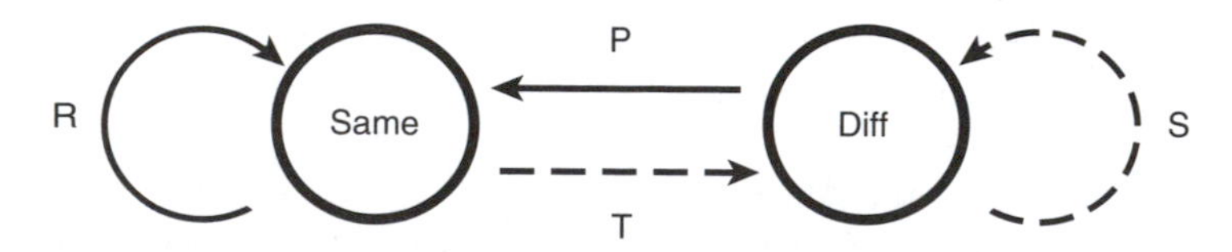

Figure 3.10 The *Pavlov* strategy described by a two-state automaton. The co-player is assumed to play *Pavlov*. The solid and dashed arrows respectively describe the transition if the player follows the *Pavlov* strategy or deviates from it. *Pavlov* is a best reply to itself if $2R > T + P$ (in the Donation game, if $b > 2c$).

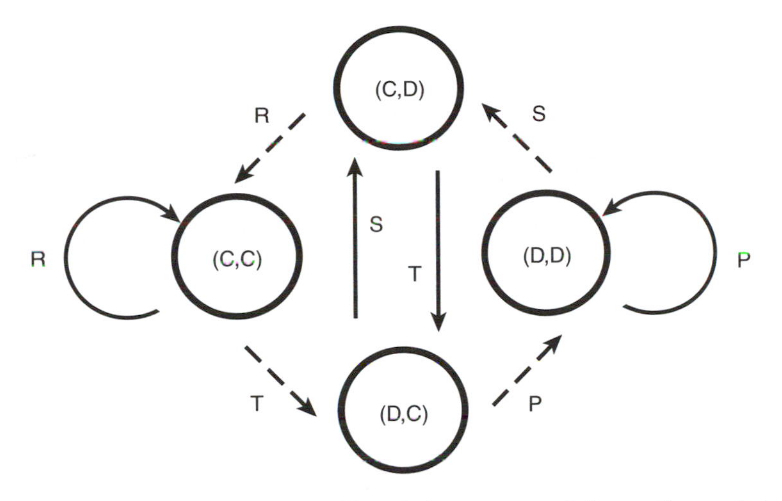

Figure 3.11 The *TFT* strategy described by a finite automaton. It is not a best reply to itself if $2R > T + S$, a condition that always holds for the Donation game.

describe the automaton (the first entry describes the move of the player, the second that of the co-player). In figure 3.11, we describe what happens when the co-player uses the strategy *TFT*. The solid arrow leaving a node describes the transition if the player, at that node, also uses the move prescribed by *TFT*, and the dashed arrow the outcome of the alternative move.

We note that *TFT* is not the best answer against itself in the Donation game. In state (C, D), the best move against a *TFT* player would be to cooperate (and to reach state (C, C)). However, *TFT* calls for one to play D, and this locks two *TFT* players into an endless cycle of unilateral defections. The payoff per round, then, is $(b - c)/2$, which is less than the payoff per round $b - c$ obtained if the node (C, C) had been reached. Of course, two *TFT* players would start out at node (C, C), and in that state *TFT* prescribes the right move. But an error leading to node (C, D) displays the fatal weakness of *TFT*. More generally, this strategy is not a best answer to itself if $2R > T + S$.

The method can be extended for all strategies where memory depends on the last two, or the last N rounds. But we shall presently see that some very simple strategies implemented by finite automata cannot be described as strategies conditioned on a prescribed number of rounds.

3.17 CONTRITE TIT FOR TAT

An interesting example for this is *CTFT* (*Contrite TFT*). Imagine that a *TFT* player who mis-implemented a C move is aware of having done wrong, and accepts meekly that the co-player, in the next round, defects in retaliation. In this case, the pointless vendetta of alternating unilateral defection can be avoided and mutual cooperation resumed. To model this, let us introduce the *standing* of a player, which

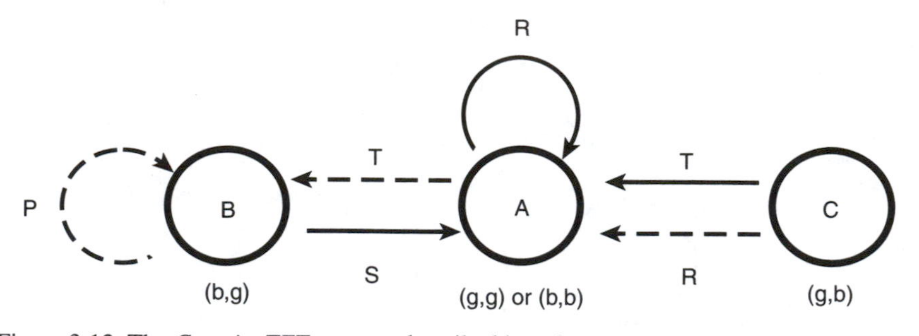

Figure 3.12 The *Contrite TFT* strategy described by a 3-state automaton. It always is a best reply to itself.

can be g or b ("good" or "bad"). Players start out in good standing and keep it until they commit an unjustified defection (i.e., until they play D while the co-player was in good standing). The good standing is regained by playing C. In any given round, a player can cooperate, commit a justified defection or an unjustified defection.

Contrite TFT is the strategy that calls for one to cooperate except when in good standing while the other player is not. This means that the player defects when provoked, but not otherwise. Thus if two *Contrite TFT* players engage in a repeated Prisoner's Dilemma game, they will always cooperate, except by mistake. After such a mistake, they will resume cooperation, and accept the co-player's retaliatory D without feeling abused.

As before, we can describe the game with a graph, and check that if the other player uses *Contrite TFT*, it is always best to also use the move prescribed by *Contrite TFT*, see figure 3.12. The nodes of the graph, i.e., the state of the game, will be A, B, and C. A corresponds to (g, g) or (b, b), B to (b, g) (the player's standing is bad and the co-player's good), and C to (g, b). We note that if one player is in state A, so is the other, whereas if one player is in state B, the other is in state C and vice versa. The rule for playing *Contrite TFT* calls for one to use C when in state A or B, and to defect (i.e., use D) in state C only. The corresponding graph shows immediately that it is best, against a *Contrite TFT* player, also to use the *Contrite TFT* rule.

The *Contrite TFT* strategy cannot be described as a memory-one strategy. Neither does it follow a rule that depends only on the outcome of a given number N of preceding rounds. Indeed, suppose that we observe a string of mutual defections in the previous N rounds. If we do not know what happened before these N rounds, we cannot say who of the two players is in good, and who is in bad standing. Hence we cannot specify what a *Contrite TFT* player ought to do in the next round.

It is interesting to compare *Contrite TFT* and *Pavlov* for the Donation game. *Contrite TFT* is always a best response to itself, *Pavlov* only if $b > 2c$. Both strategies cooperate with their like, and can easily return to mutual cooperation after an accidental error in implementation. *Contrite TFT* has the huge advantage that it is as good as *TFT* at invading a population of *AllD* players; *Pavlov*, as we have seen, is hopeless at this task, and needs a retaliatory strategy to pave the way. On the other hand, in a society dominated by *Contrite TFT*, indiscriminate altruists do just as well and hence can spread by neutral drift, eventually allowing *AllD* to invade and destroy the cooperative regime. By contrast, a society of *Pavlov* players will not allow *AllC* players to spread. As soon as the first error of implementation occurs, an *AllC* player will be exploited to the hilt.

If *Pavlov* does not fare well, i.e., if $c < b < 2c$, another strategy based on standing fares as well as *Contrite TFT*: this is *Remorse*, a strategy where a player cooperates only when in bad standing, or if both players had cooperated in the previous round. After a unilateral error, two *Remorse* players defect twice. If a *Remorse* player encounters a *Pavlov* player, both obtain an average payoff of $5(b - c)/7$ per round.

3.18 ERRORS IN PERCEPTION

Contrite TFT has its Achilles heel, too. So far, we have only considered errors in implementation. What about errors in perception? In that case, players can believe themselves to be in good standing, whereas their co-player sees them in bad standing. Two *Contrite TFT* players will, in such a situation, relentlessly inflict D upon each other, both believing that their own moves are justified defections and that their co-player's moves are not. In contrast, if an error in perception occurs between two *Pavlov* players, cooperation will be smoothly resumed after the usual mutual punishment round.

In the realm of memory-one strategies, if there is a probability ϵ to mis-implement a move, then the propensity q_R to play C after a round with outcome R is replaced by the propensity $(1 - \epsilon)q_R + \epsilon(1 - q_R)$, etc., so that the "correction term"

$$\epsilon(1 - 2q_R, 1 - 2q_S, 1 - 2q_T, 1 - 2q_P) \tag{3.77}$$

has to be added to the reaction norm (q_R, q_S, q_T, q_P). If the error affects the perception of the co-player's move (i.e., if the player confuses an R with an S, or a T with a P) then q_R turns into $(1 - \nu)q_R + \nu q_S$ etc., and the correction term is

$$\nu(q_S - q_R, q_R - q_S, q_P - q_T, q_T - q_P). \tag{3.78}$$

If the error μ affects the perception of the player's own move (i.e., a player confuses an R with a T, or an S with a P), then the correction term is

$$\mu(q_T - q_R, q_P - q_S, q_R - q_T, q_S - q_P). \tag{3.79}$$

If both types of errors in perception are admitted, then the reaction norm of *TFT*, i.e., $(1, 0, 1, 0)$, turns into $(1 - \nu, \nu, 1 - \nu, \nu)$, and *Pavlov* $(1, 0, 0, 1)$ is modified into $(1 - (\nu + \mu), \nu + \mu, \nu + \mu, 1 - (\nu + \mu))$, whereas the unconditional strategies *AllD* and *AllC* are unaffected. For $w = 1$ and the limit $\epsilon \to 0$, errors in implementation yield as payoff $(2P + 2S + T)/5$ for an S_8 player using the *Grim* strategy $(1, 0, 0, 0)$ against an S_2 player using $(0, 0, 1, 0)$, whereas errors in perceiving the opponent's move yield as payoff $(S + T)/2$, etc.

Thus it is important to consider different possibilities of errors. For instance, we might make (as in section 3.10) the plausible assumption that errors occur only if one wants to implement a C, but not if one decides to play D. In this case, the stationary distribution for an $S_8(\varepsilon)$ player against an $S_{11}(\varepsilon)$ player is $(0, 0, 1/2, 1/2)$ instead of $(1/5, 0, 2/5, 2/5)$ (see section 3.14), and thus the payoff is $b/2$ instead of $(3b - c)/5$ (up to terms in ε). The tolerant *FirmButFair* player tries vainly, every second round, to resume cooperation. The payoff for *TFT* against itself is worse now (namely 0), but the payoff in a *Pavlov* population remains unchanged. Again, a *Pavlov* population cannot be invaded if $b > 2c$.

Even among automata with only three or four inner states, there exists a bewildering number of strategies. It seems hard to figure out which one would be selected by evolution. Individual based simulations display a lot of contingencies, and offer few robust predictions. We run up against a complexity wall. On the other hand, it seems tempting to interpret the "inner states" of automata with our emotions, such as anger at being provoked, guilt at having deviated from the norm, etc.

3.19 TRIGGERS AND EQUALIZERS

The so-called *folk theorem on repeated games* is a collection of results. In the simplest setup, for two players I and II engaged in an infinitely repeated Donation game, it states that any pair $(P_{\mathrm{I}}, P_{\mathrm{II}})$ of payoff values (per round) with $0 \le P_{\mathrm{I}}, P_{\mathrm{II}} \le b - c$ can be realized by a Nash equilibrium pair of strategies. The two players simply have to follow so-called *trigger strategies*: this means playing a well-specified sequence of moves leading to $(P_{\mathrm{I}}, P_{\mathrm{II}})$, but switching to a relentless, infinite sequence of D moves as soon as the co-player deviates. It is obvious, then, that the co-player has no incentive to deviate: there is no better alternative than to follow the specified sequence of moves. In fact, any pair of payoff values can be reached such that P_{I} and P_{II} are positive and $(P_{\mathrm{I}} P_{\mathrm{II}})$ in the convex hull spanned by $(0, 0)$, $(b, -c)$, $(-c, b)$, and $(b - c, b - c)$, see figure 3.13.

This result can be extended in many ways, by considering iterations of other games (the lower bound 0 will then have to be replaced by the maximin payoff, i.e., the highest payoff that players can guarantee themselves, irrespective of their co-player's strategy), by introducing a discount on future payoffs (or allowing the iteration to stop with a positive probability), by admitting the possibility that players mis-implement their moves, etc.

The concept of a trigger strategy is often criticized on the grounds that it is too stern: it is hard to imagine that players will commit themselves forever to ruinous defection, if their co-player made a mistake just once, possibly through *force majeure*. Nevertheless, trigger strategies are an essential tool for analyzing games between rational players. In evolutionary game theory, however, trigger strategies play a less conspicuous role.

It turns out that a variant of the folk theorem can easily be displayed in the context of memory-one strategies. Indeed, there exist strategies that act as *equalizers*, in the sense that co-players always obtain the same payoff, irrespective of their strategy.

For the infinitely repeated Prisoner's Dilemma game, there exist, for every value π between P and R, memory-one strategies $\mathbf{q} = (q_R, q_S, q_T, q_P)$ such that every op-

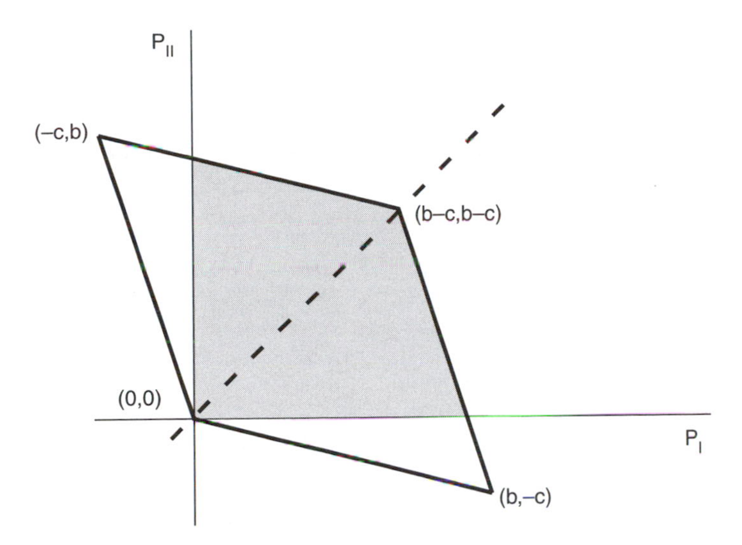

Figure 3.13 Any pair of payoff values $(P_{\mathrm{I}}, P_{\mathrm{II}})$ in the shaded region can be obtained if the two players I and II use the corresponding equalizer strategy for the infinitely repeated Prisoner's Dilemma game.

ponent obtains the long-run average payoff π against a player using such a strategy. The reaction norm **q** is given by

$$(1 - (R - \pi)a, 1 - (T - \pi)a, (\pi - S)a, (\pi - P)a), \tag{3.80}$$

where $a > 0$ is any real number such that $\frac{1}{a} \geq \max\{T - \pi, R - \pi, \pi - S, \pi - P\}$. (The condition on a guarantees that the q_i are probabilities.)

Indeed, let us denote by $p_i(n)$ the conditional probability that the player II uses the move C in round $n+1$, given that the n-th round resulted in outcome i for player I; and let $s_i(n)$ be the probability of that outcome. By conditioning on round n, we see that $s_R(n+1)$ is given by

$$\begin{aligned} &s_R(n)p_R(n)[1 - (R - \pi)a] + s_S(n)p_S(n)[1 - (T - \pi)a] \\ &\quad + s_T(n)p_T(n)(\pi - S)a + s_P(n)p_P(n)(\pi - P)a. \end{aligned} \tag{3.81}$$

Similarly, $s_S(n+1)$ is given by

$$\begin{aligned} &s_R(n)(1 - p_R(n))[1 - (R - \pi)a] + s_S(n)(1 - p_S(n))[1 - (T - \pi)a] \\ &\quad + s_T(n)(1 - p_T(n))(\pi - S)a + s_P(n)(1 - p_P(n))(\pi - P)a. \end{aligned} \tag{3.82}$$

Summing these equations yields the probability that player I chooses move C in round $n+1$, namely $s_R(n+1) + s_S(n+1)$. It is given by

$$s_R(n)[1-(R-\pi)a]+s_S(n)[1-(T-\pi)a]+s_T(n)(\pi-S)a+s_P(n)(\pi-P)a. \tag{3.83}$$

Hence

$$s_R(n) + s_S(n) - s_R(n+1) - s_S(n+1) \tag{3.84}$$

is given, up to the factor a, by

$$Rs_R(n)+Ss_T(n)+Ts_S(n)+Ps_P(n)-\pi[s_R(n)+s_S(n)+s_T(n)+s_P(n)]. \quad (3.85)$$

Since the $s_i(n)$ sum up to 1, this is just $A'(n)-\pi$, where $A'(n)$ is player II's payoff in the n-th round. (We must bear in mind that one player's S is the other player's T.) Summing up for $n=0,\ldots,N$ and dividing by $N+1$, we obtain

$$\frac{s_R(0)+s_S(0)-s_R(N)-s_S(N)}{a(N+1)}=\frac{A'(0)+\cdots+A'(N)}{N+1}-\pi, \quad (3.86)$$

which yields

$$\lim_{N\to\infty}\frac{A'(0)+\cdots+A'(N)}{N+1}=\pi. \quad (3.87)$$

3.20 THE ALTERNATING PRISONER'S DILEMMA

In many real-life instances of direct reciprocity, the two players alternate in their roles of donor and recipient, whereas in most of the literature, and in our treatment so far, the two players decide simultaneously. Usually, this assumption is of small effect. But in some situations, important differences can arise.

In an alternating Prisoner's Dilemma (or the alternating Donation game), to cooperate means to play C when it is one's turn to do so. This can affect strategies and payoffs. For instance, if two *TFT* players engage in an iterated Prisoner's Dilemma of the usual, simultaneous kind, and if one player defects by mistake, both players will subsequently play C and D in turn. In the alternating Prisoner's Dilemma game, if a unilateral defection occurs by mistake, the result will be a sequence of mutual defections: both players play D, see figure 3.14. The average payoff will be the same as in the simultaneous case, and in fact, the interplay between *TFT*, *AllC*, and *AllD* remains unchanged. But if two *Pavlov* players, for instance, are matched against each other, the outcome is very different. A mistaken D is answered by a D, which elicits a C, which is followed by a D in turn. Thus each player, after the erroneous defection, keeps playing two D's and one C periodically. With probability 2/3, the

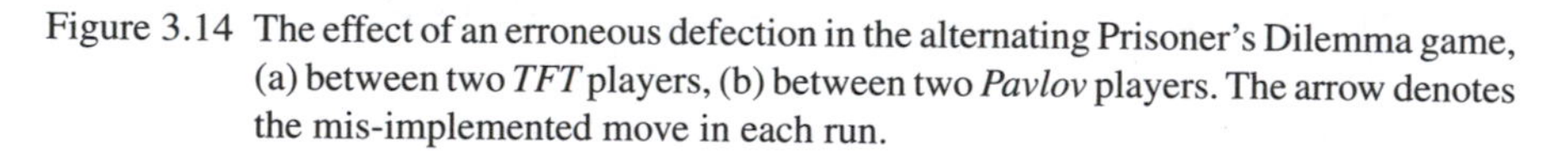

Figure 3.14 The effect of an erroneous defection in the alternating Prisoner's Dilemma game, (a) between two *TFT* players, (b) between two *Pavlov* players. The arrow denotes the mis-implemented move in each run.

Table 3.2 The Alternating Donation Game with Errors in Implementation.

$*$	S_0	S_1	S_2	S_3	S_4	S_5	S_6	S_7	S_8	S_9	S_{10}	S_{11}	S_{12}	S_{13}	S_{14}	S_{15}
S_0	$\mathbf{0}$	$\frac{b}{2}$	0	$\frac{b}{2}$	$\frac{b}{3}$	b	$\frac{b}{2}$	b	0	$\frac{b}{2}$	0	$\frac{b}{2}$	$\frac{b}{2}$	b	$\frac{2b}{3}$	b
S_1	$-\frac{c}{2}$	$\mathbf{\frac{b-c}{3}}$	$\frac{b-c}{2}$	$\frac{b-c}{2}$	$\frac{b-c}{3}$	b	b	b	$-\frac{c}{2}$	$\frac{b-c}{3}$	$\frac{b-c}{2}$	$\frac{b-c}{2}$	$\frac{2b-c}{4}$	b	b	b
S_2	0	$\frac{b-c}{2}$	$\mathbf{0}$	$\frac{b-c}{2}$	0	$\frac{b-c}{2}$	0	$\frac{b-c}{2}$	0	$\frac{2b-c}{3}$	0	$\frac{2b-c}{3}$	$\frac{2b-c}{4}$	$\frac{2b-c}{2}$	$\frac{2b-c}{3}$	$\frac{2b-c}{2}$
S_3	$-\frac{c}{2}$	$\frac{b-c}{2}$	$\frac{b-c}{2}$	$\mathbf{\frac{b-c}{2}}$	$-\frac{c}{2}$	$\frac{b-c}{2}$	$\frac{b-c}{2}$	$\frac{b-c}{2}$	$-\frac{c}{2}$	$\frac{b-c}{2}$	$\frac{b-c}{2}$	$\frac{b-c}{2}$	$\frac{b-c}{2}$	$\frac{2b-c}{2}$	$\frac{2b-c}{2}$	$\frac{2b-c}{2}$
S_4	$-\frac{c}{3}$	$\frac{b-c}{3}$	0	$\frac{b}{2}$	$\mathbf{\frac{b-c}{3}}$	$\frac{2b-c}{3}$	$\frac{2b}{3}$	b	$-\frac{c}{3}$	$\frac{2b-c}{5}$	0	$\frac{b}{2}$	$\frac{3b-c}{6}$	b	$\frac{2b}{3}$	b
S_5	$-c$	$-c$	$\frac{b-c}{2}$	$\frac{b-c}{2}$	$\frac{b-2c}{3}$	$\mathbf{\frac{b-c}{2}}$	b	b	$-c$	$-c$	$\frac{b-c}{2}$	$\frac{b-c}{2}$	$\frac{b-c}{2}$	$\frac{2b-c}{3}$	b	b
S_6	$-\frac{c}{2}$	$-c$	0	$\frac{b-c}{2}$	$-\frac{2c}{3}$	$-c$	$\mathbf{\frac{b-c}{2}}$	$\frac{2b-2c}{3}$	$-\frac{c}{3}$	$\frac{b-c}{2}$	0	$\frac{2b-c}{3}$	$\frac{b-c}{2}$	$\frac{4b-3c}{5}$	$\frac{4b-2c}{5}$	$\frac{2b-c}{2}$
S_7	$-c$	$-c$	$\frac{b-c}{2}$	$\frac{b-c}{2}$	$-c$	$-c$	$\frac{2b-2c}{3}$	$\mathbf{\frac{2b-2c}{3}}$	$-c$	$-c$	$\frac{b-c}{2}$	$\frac{b-c}{2}$	$\frac{2b-3c}{4}$	$\frac{2b-2c}{3}$	$\frac{2b-c}{2}$	$\frac{2b-c}{2}$
S_8	0	$\frac{b}{2}$	0	$\frac{b}{2}$	$\frac{b}{3}$	b	$\frac{b}{3}$	b	$\mathbf{0}$	$\frac{3b-c}{5}$	$\frac{b-c}{3}$	$\frac{2b-c}{3}$	$\frac{3b-c}{6}$	$\frac{3b-c}{3}$	$\frac{2b-c}{3}$	$\frac{3b-c}{3}$
S_9	$-\frac{c}{2}$	$\frac{b-c}{3}$	$\frac{b-2c}{3}$	$\frac{b-c}{2}$	$\frac{b-2c}{5}$	b	$\frac{b-c}{2}$	b	$\frac{b-3c}{5}$	$\mathbf{\frac{b-c}{2}}$	$b-c$	$b-c$	$\frac{b-c}{2}$	$\frac{3b-c}{3}$	$\frac{3b-2c}{3}$	$\frac{2b-c}{2}$
S_{10}	0	$\frac{b-c}{2}$	0	$\frac{b-c}{2}$	0	$\frac{b-c}{2}$	0	$\frac{b-c}{2}$	$\frac{b-c}{3}$	$b-c$	$\mathbf{\frac{b-c}{2}}$	$b-c$	$\frac{b-c}{2}$	$b-c$	$\frac{2b-2c}{3}$	$b-c$
S_{11}	$-\frac{c}{2}$	$\frac{b-c}{2}$	$\frac{b-2c}{3}$	$\frac{b-c}{2}$	$-\frac{c}{2}$	$\frac{b-c}{2}$	$\frac{b-2c}{3}$	$\frac{b-c}{2}$	$\frac{b-2c}{3}$	$b-c$	$b-c$	$\mathbf{b-c}$	$\frac{2b-3c}{4}$	$b-c$	$b-c$	$b-c$
S_{12}	$-\frac{c}{2}$	$\frac{b-2c}{4}$	$\frac{b-2c}{4}$	$\frac{b-c}{2}$	$\frac{b-3c}{6}$	$\frac{b-c}{2}$	$\frac{b-c}{2}$	$\frac{3b-2c}{4}$	$\frac{b-3c}{6}$	$\frac{b-c}{2}$	$\frac{b-c}{2}$	$\frac{3b-2c}{4}$	$\mathbf{\frac{b-c}{2}}$	$\frac{5b-3c}{6}$	$\frac{5b-3c}{6}$	$\frac{2b-c}{2}$
S_{13}	$-c$	$-c$	$\frac{b-2c}{2}$	$\frac{b-2c}{2}$	$-c$	$\frac{b-2c}{3}$	$\frac{3b-4c}{5}$	$\frac{2b-2c}{3}$	$\frac{b-3c}{3}$	$\frac{b-3c}{3}$	$b-c$	$b-c$	$\frac{3b-5c}{6}$	$\mathbf{\frac{2b-2c}{3}}$	$\frac{3b-2c}{3}$	$\frac{3b-2c}{3}$
S_{14}	$-\frac{2c}{3}$	$-c$	$\frac{b-2c}{3}$	$\frac{b-2c}{2}$	$-\frac{2c}{3}$	$-c$	$\frac{2b-4c}{5}$	$\frac{b-2c}{2}$	$\frac{b-2c}{3}$	$\frac{2b-3c}{3}$	$\frac{2b-2c}{3}$	$b-c$	$\frac{3b-5c}{6}$	$\frac{2b-3c}{3}$	$\mathbf{b-c}$	$b-c$
S_{15}	$-c$	$-c$	$\frac{b-2c}{2}$	$\frac{b-2c}{2}$	$-c$	$-c$	$\frac{b-2c}{2}$	$\frac{b-2c}{2}$	$\frac{b-3c}{3}$	$\frac{b-2c}{2}$	$b-c$	$b-c$	$\frac{b-2c}{2}$	$\frac{2b-3c}{3}$	$b-c$	$\mathbf{b-c}$

next mistake will not affect this regime. Only with probability 1/3 will it redress the game to a run of mutual cooperation. The average payoff is $(b-c)/2$ in the infinitely repeated case ($w=1$).

If players alternate in being the potential donor, then two consecutive rounds of the alternating game correspond to one round of the simultaneous game. Let us assume that the memory of each player covers the previous two rounds (i.e., one decision by each player on whether to donate or not). The outcomes will be denoted in the obvious way by R, S, T, and P, and the strategies for the infinitely iterated alternating game by the propensities q_R, q_S, etc., to cooperate after outcome R, S, etc. The transition probabilities for a (q_R, q_S, q_T, q_P) player against a (q'_R, q'_S, q'_T, q'_P) player are given by the matrix

$$Q = \begin{pmatrix} q_R q'_R & q_R(1-q'_R) & (1-q_R)q'_S & (1-q_R)(1-q'_S) \\ q_S q'_T & q_S(1-q'_T) & (1-q_S)q'_P & (1-q_S)(1-q'_P) \\ q_T q'_R & q_T(1-q'_R) & (1-q_T)q'_S & (1-q_T)(1-q'_S) \\ q_P q'_T & q_P(1-q'_T) & (1-q_P)q'_P & (1-q_P)(1-q'_P) \end{pmatrix}, \tag{3.88}$$

which is quite different from matrix (3.67). The payoff can be computed as before. It turns out that in the alternating Prisoner's Dilemma, *Pavlov* loses much of its appeal. As table 3.2 shows, its place is taken up by *Firm But Fair*, with reaction norm $S_{11}=(1, 0, 1, 1)$. This strategy is error-correcting and achieves the highest payoff against itself, namely $b-c$, just as S_{14} and S_{15} do. But in contrast to these latter two strategies, *Firm But Fair* cannot be invaded by other strategies, such as *AllD*, as long as $b>2c$. However, the strategy S_{14} can enter by neutral drift. On the other hand, *AllD* $=S_0$ can always be invaded by S_8 and S_{10}, which in turn can be invaded by S_{11}. If we consider only errors in implementing a cooperative move, we see as in section 3.14 that *AllD* can be subverted by many strategies through neutral drift, and that among these, S_2, S_6, S_{10}, and S_{14} give way to *Firm But Fair*.

If we restrict attention to reactive strategies, for which $q_R=q_T=p$ and $q_S=q_P=q$, we find that the payoffs for the donation game are exactly as for the simultaneous game, although the sequence of moves can be quite different (as we have seen in the instance of two *TFT* players). Again, *Generous TFT* emerges as the winner. Within the realm of strategies given by finite automata, *Contrite TFT* is as good in the alternating as in the simultaneous Prisoner's Dilemma, and as vulnerable to errors in perception.

3.21 REFERENCES

Trivers (1971) introduced reciprocal altruism as a major factor in the evolution of cooperation. The use of the repeated Prisoner's Dilemma was taken up by Axelrod and Hamilton (1981) and started a huge wave of investigations, see Axelrod (1984), Axelrod and Dion (1988), Trivers (2006), and Kendall, Yao, and Chong (2007). But this had been preceded by many theoretical and empirical investigations, see e.g., Rapoport and Chammah (1965). For various views on the importance of Tit for Tat among non-human players, we refer to Dugatkin (1997), Milinski (1987), and Hammerstein (2003). The Chicken game, also known as Hawk-Dove game, played an essential role in the early development of evolutionary game theory, see Maynard Smith (1982). In the form of the Snowdrift game, its relevance for cooperation was highlighted by Sugden (1986). The treatment in sections 3.2 to 3.4 on repeated Donation games closely follows Brandt and Sigmund (2006), and that in sections 3.5 to 3.8 on reactive strategies follows Nowak and Sigmund (1990). Adaptive dynamics was introduced in Nowak and Sigmund (1989) and Hofbauer and Sigmund (1990), and has been greatly developed since, see Dieckmann and Law (1996) or Dieckmann and Metz (2009). Molander (1985) introduced *Generous TFT*, see also Nowak and Sigmund (1992). Memory-one strategies are discussed in Lindgren (1991) and Nowak, Sigmund, and El-Sedy (1995), see Hilbe (2008) for an exhaustive treatment. The success of the *Pavlov* strategy was first noted in Kraines and Kraines (1989) and Fudenberg and Maskin (1990), its role in evolutionary dynamics was pointed out by Nowak and Sigmund (1993) and, for finite populations, by Imhof, Fudenberg, and Nowak (2007). May (1987), Boyd (1989), Bendor and Swistak (1995), and Sheratt and Roberts (2001) stress the role of errors and occasional defections in stabilizing cooperation. For strategies implemented by finite automata, see Aumann (1981), Rubinstein (1986), Abreu and Rubinstein (1988), Banks and Sundaram (1990), Binmore and Samuelson (1992), and Leimar (1997). The role of errors was stressed by Boyd (1989), see also May (1987). *Contrite TFT* was proposed by Sugden (1986), see also Boerlijst, Nowak, and Sigmund (1997a). For economic experiments, see e.g. Milinski and Wedekind (1998), Kollock (1993), and Kagel and Roth (1995). From the vast literature on the folk theorem for repeated games, we refer particularly to Selten and Hammerstein (1984) and Fudenberg and Maskin (1986). The section on equalizer strategies follows Boerlijst, Nowak, and Sigmund (1997b). The alternating Prisoner's Dilemma is studied in Nowak and Sigmund (1994), Frean (1994), and Neill (2001). For the important topic of continuous investment levels, see Roberts and Sherratt (1998), Wahl and Nowak (1999), and Killingback and Doebeli (2002). Fishman, Lotem, and Stone (2001), Johnson, Stopka, and Bell (2002), and McNamara, Barta, and Houston (2004) stress the importance of variability in behavior.

Chapter Four

Indirect Reciprocity: The Role of Reputation

4.1 INDIRECT RECIPROCITY

In the previous chapter, we have investigated strategies for playing the Donation game repeatedly against *the same* co-player. In this chapter, we assume that the same game is played repeatedly, but always against *another* co-player. Third parties have to return the helpful action. This introduces major differences between direct and indirect reciprocation, as we shall presently see.

The *Tit for Tat* strategy, which played such a basic role in the previous chapter, discriminates according to the outcome of the previous round. In direct reciprocity, what happens to a player is caused by what the co-player does. But in the context of indirect reciprocity, this is no longer the case: the two players had different partners in their previous rounds. Accordingly, there are two different ways of reciprocating. Players can either base their decision (to donate or not) on what happened to them in the previous round; or else, they can base their decision on what their co-player did in the previous round.

Roughly speaking, players can either be affected by a diffuse feeling of indebtedness—"Somebody helped me, I feel elated and therefore will help another person,"—or else, they can be moved by a feeling of appreciation—"My co-player acted graciously, not to me but to another person, and I will now help my co-player in return."

In one case, A gives to B and therefore B gives to C. We may view this as *misguided* reciprocation: the return is addressed to C instead of A. In the other case, A gives to B and therefore C gives to A, see figure 4.1. This may be termed *vicarious* reciprocation: A receives the deserved return, not from B but from a third party C. In one case, the reciprocator received a benefit, and in the next round expresses gratitude to a person who did not help him. In the other case, the reciprocator rewards a benefactor—but for an action that benefitted someone else. Vicarious and misguided reciprocity are also called downstream and upstream reciprocity. In the former case, players incur costs in the hope of recouping them later; in the other case, they can afford the costs because of previous benefits.

Interestingly, both factors seem to show up in economic experiments. But in the theoretical models considered so far, vicarious reciprocity, i.e., rewarding, works fairly well, whereas misguided reciprocity, i.e., thanking, seems much more difficult to explain.

Another difference between direct and indirect reciprocation is that two players engaged in direct reciprocation experience the same number of rounds in parallel. (They do so even if they alternate in the roles of the donor or the recipient.) By

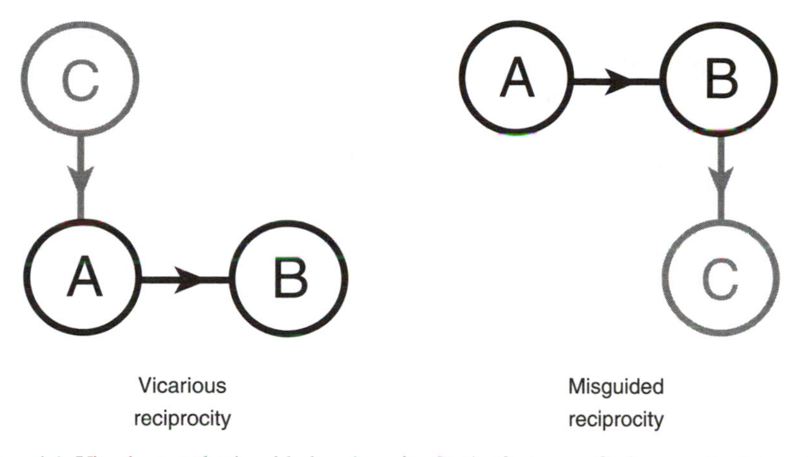

Figure 4.1 Vicarious and misguided reciprocity. In the first case, C observes the interaction between A and B and then decides whether or not to help A. In the other case, whether B received some help from A determines whether B helps C in turn.

contrast, the histories of two players interacting via indirect reciprocity intersect only once, and thus the players have a different numbering of their rounds: a donor in the second round may be matched with a recipient who has reached the fifth round. Some models assume that the players are synchronized, starting out at the same time, experiencing their interactions in the same rhythm and updating their strategies at the same signal, but this is a contrived feature and will therefore be omitted.

A high value for w, the probability of another round, is less plausible with indirect than with direct reciprocity, since in a realistically small population, players experiencing many rounds would necessarily have to interact numerous times with the same partner, and hence be engaged in direct reciprocity. But in the following model of indirect reciprocity, the limiting case $w = 1$ does not alter the outcome.

4.2 THE GOOD, THE BAD, AND THE RECIPROCATOR

We will consider a *continuous entry* model: this means that players enter a large population one by one, interact asynchronously with different players at random times, update their strategy occasionally, and eventually exit. Since we assume that the population is large, its composition will evolve so slowly that we may assume it does not change from one round to the next. We could assume that in any given interaction between two players, one is randomly assigned the role of donor, and the other that of recipient; but in order to simplify the formulas, we shall assume that in any given round, a player is both donor and recipient. The player's donor and recipient are different, but we can nevertheless describe the payoff, in each round, by the matrix (3.1) of the Donation game.

We consider the case of vicarious reciprocation. With x and y we denote the frequencies of the unconditional strategies *AllC* and *AllD*, and with z that of the reciprocators. We begin by considering the simplest type of reciprocator, the one corresponding to *Tit for Tat* in direct reciprocity. Such players refuse to help if they know that their current recipient refused to help in the previous round. With q we denote the probability that a player knows (either through direct observation or via gossip) what the randomly assigned recipient did in the previous round. To begin with, we posit that reciprocators are *trustful* in the sense that if they have no information, they assume that their recipient gave help previously.

As in the case of direct reciprocity, we allow for error, and denote by ϵ the probability of not implementing an intended donation. This can be due to a mistake, but it could also be due to external circumstances, for instance a momentary lack of resources. We assume that an intended refusal will always be carried out.

Let h denote the frequency of players with a good reputation (i.e., having given help in their previous round). Since the population is large, this frequency remains unchanged between two consecutive rounds. Writing $\bar{\epsilon} := 1 - \epsilon$, we obtain

$$h = \bar{\epsilon}[x + z(1 - q + qh)]. \tag{4.1}$$

Indeed, players intend to donate if they are *AllC* players (probability x) or if they are reciprocators (probability z), who either know nothing of their co-player's reputation (probability $1-q$) or know the reputation (probability q), which is good (probability h). The probability for intending to donate (the term in square brackets) has to be multiplied by $\bar{\epsilon}$ (the probability not to commit an error). Hence

$$h = \frac{\bar{\epsilon}(x + (1 - q)z)}{1 - \bar{\epsilon}qz}. \tag{4.2}$$

The payoff in round n (with $n \geq 1$) for an *AllC* player is

$$P_x(n) = -c\bar{\epsilon} + b\bar{\epsilon}[x + z(1 - q + \bar{\epsilon}q)]. \tag{4.3}$$

Indeed, such a player always intends to donate, at a cost $-c$ (this succeeds with probability $\bar{\epsilon}$). On the other hand, the player is the object of an intended donation if the co-player in the role of the potential donor is either an unconditional cooperator (probability x) or a reciprocator (probability z) who either does not know the player's reputation (probability $1-q$) or else knows the reputation (probability q), which is good (probability $\bar{\epsilon}$). (This is clear because the reputation can only be bad if the player, an *AllC* player, made a mistake in the previous round.) The benefit resulting from an intended donation is $b\bar{\epsilon}$, because the donation can fail with probability ϵ.

Similarly, the payoff for a defector is

$$P_y(n) = b\bar{\epsilon}[x + (1 - q)z], \tag{4.4}$$

and for a reciprocator, whom we call A, it is

$$P_z(n) = -c\bar{\epsilon}(1 - q + qh) + b\bar{\epsilon}[x + z[1 - q + \bar{\epsilon}q(1 - q + qh)]]. \tag{4.5}$$

The second term in the sum is (up to the expected benefit $b\bar{\epsilon}$) the probability that the co-player intends to make a donation to player A. This happens either if the co-player is an *AllC* player (probability x), or if the co-player is a reciprocator (probability z) who either does not know the reputation of player A (probability $1 - q$), or else knows the reputation (probability q), and this reputation is good. The reputation of A is good if in the previous round, player A intended to donate (either because A did not know the co-player's reputation or else because that reputation was good, probability h), and if A, moreover, succeeded in the intended donation (probability $\bar{\epsilon}$).

A straightforward computation shows that

$$P_z(n) - P_y(n) = [P_x(n) - P_y(n)](1 - q + qh). \tag{4.6}$$

The same relation holds for the initial round (numbered 0), although the payoffs for the initial round are slightly different: $P_x(0) = -c\bar{\epsilon} + b\bar{\epsilon}(x+z)$, $P_y(0) = b\bar{\epsilon}(x+z)$, and $P_z(0) = -c\bar{\epsilon}[1-q+qh] + b\bar{\epsilon}(x+z)$. Hence the payoff values *per round* P_x, P_y, and P_z, given as in section 3.2, also satisfy

$$P_z - P_y = [P_x - P_y](1 - q + qh). \tag{4.7}$$

Clearly $P_x(n) - P_y(n) = \bar{\epsilon}(-c + b\bar{\epsilon}qz)$ (for $n \geq 1$) and $P_x(0) - P_y(0) = -c\bar{\epsilon}$. Thus the payoff values per round satisfy

$$P_x - P_y = \bar{\epsilon}(-c + wb\bar{\epsilon}qz), \tag{4.8}$$

which also holds if $w = 1$.

4.3 REPLICATOR DYNAMICS

In order to study the replicator equation $\dot{x} = x(P_x - \bar{P})$ etc., we can subtract P_y from each payoff and divide by the factor $\bar{\epsilon}$. By abuse of notation, the resulting expressions will again be denoted by P_x, P_y, and P_z. Clearly,

$$P_x = f, \qquad P_y = 0, \qquad P_z = f(1 - q + qh), \tag{4.9}$$

where

$$f = -c + wb\bar{\epsilon}qz. \tag{4.10}$$

Let us first consider the replicator equation obtained by omitting the common factor f (i.e., by replacing f by 1). The term h is given by equation (4.2), i.e., a fraction whose denominator $1 - \bar{\epsilon}qz$ is always positive. We can multiply the right hand sides of the replicator equation by this positive term without altering the orbits (only the velocities of the solutions are changed, but not their paths). This yields a replicator equation with

$$P_x = 1 - \bar{\epsilon}qz, \quad P_y = 0, \quad P_z = 1 - q + \bar{\epsilon}qx. \tag{4.11}$$

If $q < 1$ and $\epsilon > 0$, we have $0 = P_y < P_z < P_x$ and hence all orbits in S_3 converge to $x = 1$, with the exception of the edge $x = 0$, which leads to $z = 1$. An invariant of motion is given by $V = zx^{q-1}y^{-\epsilon q}$, as can be seen by a straightforward computation.

If $\epsilon = 0$ (no errors), the edge $y = 0$ consists of fixed points and the invariant of motion is $V = zx^{q-1}$. If $q = 1$ (full information about the co-players), the edge $x = 0$ consists of fixed points and the invariant of motion is $V = zy^{-\epsilon}$.

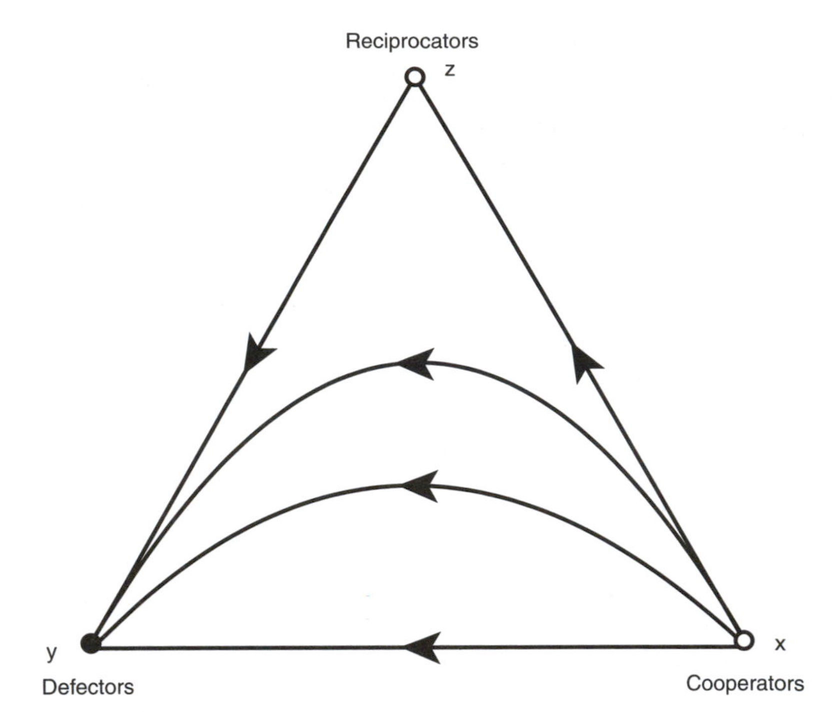

Figure 4.2 Defectors always win if players do not have enough information about their co-players.

Let us now consider the replicator dynamics given by equations (4.9) (i.e., including the factor f).

If $q < c/wb\bar{\epsilon}$, then f is negative for all values of z between 0 and 1, and hence on the whole state simplex S_3. Multiplication with f corresponds thus to a time-reversal. This means that the *AllC* players are dominated by both the reciprocators and the defectors, while the reciprocators are dominated by the defectors. All orbits in the interior of the simplex lead from $x = 1$ (*AllC* players only) to $y = 1$ (*AllD* players only). Hence, if the probability q to know the co-players' past is too small (i.e., if there is not much scope for reputation), then cooperation cannot evolve, see figure 4.2. If $q > c/wb\bar{\epsilon}$, then the line $z = c/wbq\bar{\epsilon}$ intersects the interior of the simplex S_3 and defines a segment of rest points. Indeed, on that line, $0 = P_y = P_x = P_z$. These rest points are all Nash equilibria. In the simplex S_3, all orbits lie on the same curves as when f is replaced by 1, so that $zx^{q-1}y^{-\epsilon q}$ is constant, but in contrast to the previous situation, the orientation has *not* changed in the region with $z > c/wbq\bar{\epsilon}$, see figure 4.3.

This means in particular that the mixture of *AllC* players and reciprocators given by $z = c/wbq\bar{\epsilon}$ and $y = 0$ corresponds to a rest point of the replicator dynamics. A cooperative population consisting of these two types of altruists (some conditional and some not) exists, if the average level of information within the population is sufficiently high. We note that this equilibrium is stable. However, it is not asymptotically stable, since it belongs to a segment of rest points.

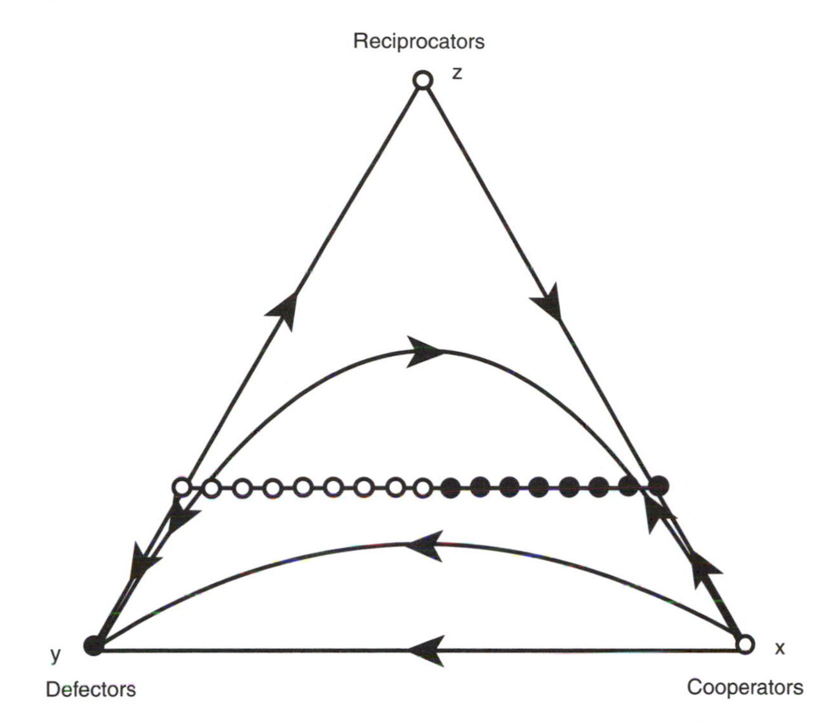

Figure 4.3 If information is sufficiently large, defectors, cooperators, and reciprocators coexist. The horizontal line of fixed points includes a stable mixture of *AllC* and reciprocators. But random shocks can lead, in the long run, to a population consisting only of defectors. (Under plausible parameter values, the region of stable fixed points, corresponding to the filled circles, is much smaller than shown here.)

The dynamic behavior in the vicinity of the horizontal line of Nash equilibria is interesting. One part of the segment is transversely stable, in the sense that perturbations away from the segment are counteracted by the dynamics. In the other part, perturbations are amplified by the dynamics. A small deviation to higher z values will lead, first to an increase and then to a decrease of reciprocators, and eventually back to the stable part of the segment. By contrast, a small deviation to lower z values leads, in the unstable part of the line of rest points, to the fixation of defectors.

In the limiting case $\epsilon = 0$ (no errors), the edge $y = 0$ consists of rest points, of which those with $z \geq c/wbq$ are Nash equilibria. The line with $z = c/wbq$ consists of rest points too. Below this line, all orbits converge to $y = 1$. Above the line, each orbit converges to a Nash equilibrium on $y = 0$, see figure 4.4. In the limiting case $q = 1$ (full information), the edge $x = 0$ consists of rest points, of which those with $z < c/wb\bar{\epsilon}$ are Nash equilibria. The line with $z = c/wb\bar{\epsilon}$ consists of rest points that are all stable. Hence the dynamics is as shown in figure 4.5.

If $q = 1$ and $\epsilon = 0$ both hold, the edges $x = 0$ and $y = 0$ both consist of rest points. In the interior of S_3, all orbits remain on parallels to the $z = 0$ edge. Those with $z > c/wb$ point from left to right (the defectors vanish), while those with $z < c/wb$

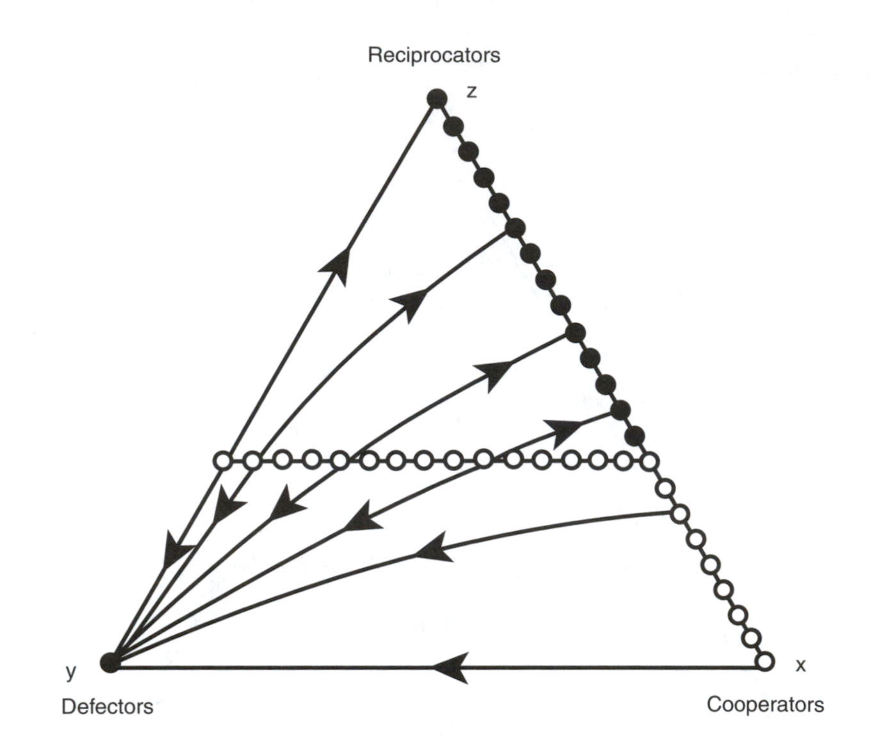

Figure 4.4 The replicator dynamics if the probability ϵ of errors in implementation is 0.

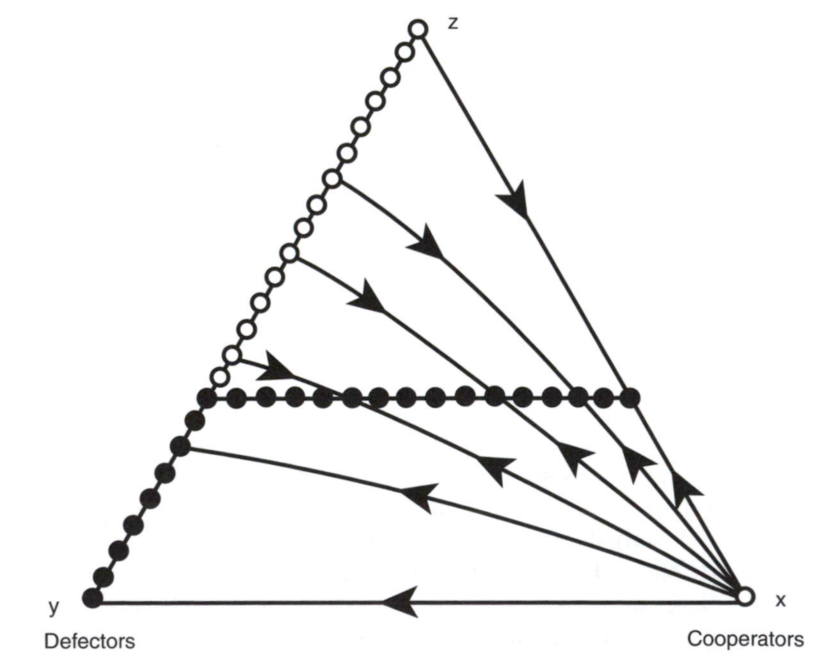

Figure 4.5 The replicator dynamics if q, the probability to know the co-player's reputation, is 1.

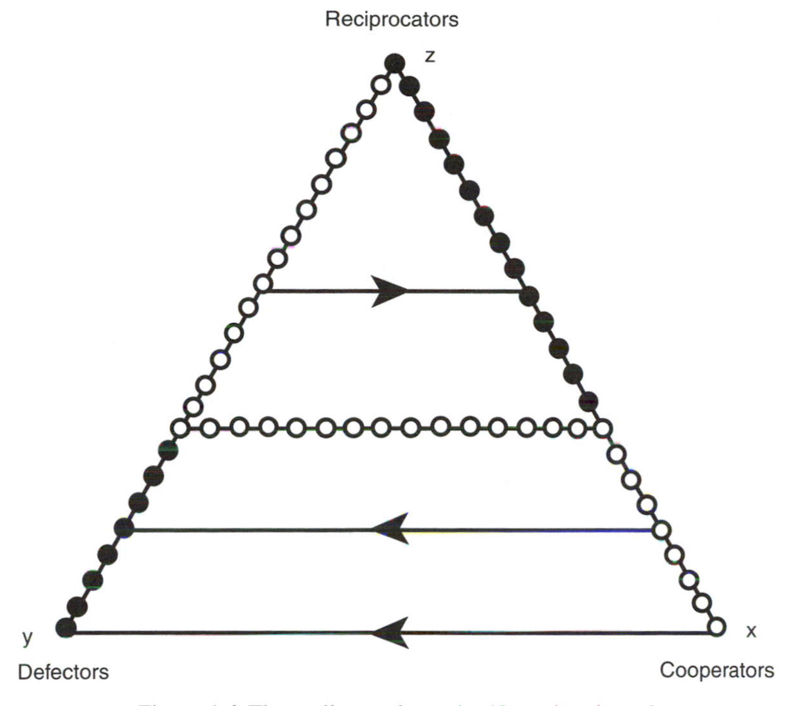

Figure 4.6 The replicator dynamics if $q = 1$ and $\epsilon = 0$.

point from right to left (the undiscriminating altruists vanish). Again, the horizontal line $z = c/wb$ consists of Nash equilibria. The dynamics is shown in figure 4.6.

Finally, let us briefly consider the case of misguided reciprocity. In this case, reciprocators decide to give help whenever they have received support in the previous round. If we denote by h the probability that a player has received support in the previous round, we see that $h = \bar{\epsilon}(x + hz)$.

In round n the payoff values for *AllC* players, *AllD* players, and reciprocators are $P_x(n) = -c\bar{\epsilon} + hb$, $P_y(n) = hb$, and $P_z(n) = -ch\bar{\epsilon} + hb$. If we assume that a reciprocator always donates in the first round, we get $P_z(0) = -c\bar{\epsilon} + bh$. Again normalizing the total payoff values such that $P_y = 0$, we obtain, up to the factor $(1 - w)^{-1}$,

$$P_x = -c\bar{\epsilon}, \qquad P_z = P_x[1 - w(1 - h)]. \tag{4.12}$$

The dynamics looks as in figure 4.2: the defector's vertex, given by $y = 1$, is a global attractor. This still holds if the error rates are modified, or if one assumes that the reciprocators defect in the first round, etc. In particular, letting $\epsilon \to 0$ or $w \to 1$ changes nothing. It is all the more surprising that some experiments (and, indeed, everyday introspection) show that indirect reciprocation based on a misdirected feeling of gratitude is not rare.

4.4 TO TRUST OR NOT TO TRUST

So far, we have assumed that reciprocators are *trustful* in the sense that by default, they assume that their co-player deserves to be helped. But the question what to do with co-players about whom nothing is known has no obvious answer. Let us therefore also introduce reciprocators who are *suspicious* and refuse to help co-players about whose antecedents they know nothing.

We thus assume that in addition to the indiscriminate altruists and defectors (with frequencies x and y) there are both trustful and suspicious reciprocators (with frequencies z and ζ). We again allow that with probability ϵ, an intended donation is not carried out (whereas intended refusals always are). In equations (4.13) to (4.17), the probability of an intended donation is always multiplied by the probability $\bar{\epsilon} = 1-\epsilon$.

The frequency of players with good reputation (those having given in their previous round) satisfies

$$h = \bar{\epsilon}[x + \zeta q h + z(1 - q + qh)]. \tag{4.13}$$

(The term $1 - q - qh$ is the probability that z players, in their previous round, have acquired a good reputation.) The payoffs in round $n \geq 1$ are

$$P_x(n) = -c\bar{\epsilon} + b\bar{\epsilon}[x + z(1 - q + q\bar{\epsilon}) + \zeta q\bar{\epsilon}], \tag{4.14}$$

$$P_y(n) = b\bar{\epsilon}[x + (1 - q)z], \tag{4.15}$$

$$P_z(n) = -c\bar{\epsilon}(1 - q + qh) + b\,\bar{\epsilon}[x + z(1 - q + q\bar{\epsilon}(1 - q + qh)) + \zeta q\bar{\epsilon}(1 - q + qh)], \tag{4.16}$$

$$P_\zeta(n) = -c\,\bar{\epsilon}qh + b\bar{\epsilon}[x + z(1 - q + q\bar{\epsilon}qh) + \zeta q\bar{\epsilon}qh]. \tag{4.17}$$

Hence

$$P_x(n) - P_y(n) = -c\bar{\epsilon} + b\,\bar{\epsilon}^2 q(z + \zeta), \tag{4.18}$$

$$P_z(n) - P_y(n) = (1 - q + qh)[P_x(n) - P_y(n)], \tag{4.19}$$

and

$$P_\zeta(n) - P_y(n) = qh[P_x(n) - P_y(n)]. \tag{4.20}$$

It is easy to see that these last three relations also hold for the initial round, i.e., for $n=0$. Thus we obtain for the total payoffs per round:

$$P_z - P_y = (P_x - P_y)(1 - q + qh) \tag{4.21}$$

and

$$P_\zeta - P_y = (P_x - P_y)qh. \tag{4.22}$$

Moreover,

$$P_x - P_y = \bar{\epsilon}[-c + bwq\bar{\epsilon}(z + \zeta)] := f. \tag{4.23}$$

Let us consider the replicator equations $\dot{x} = x(P_x - \bar{P})$, etc. We subtract P_y from all payoff terms, and first analyze the equation obtained by replacing the common factor f with 1. Thus we consider the replicator equation with

$$P_x = 1, \qquad P_y = 0, \qquad P_z = 1 - q + qh, \qquad P_\zeta = qh. \tag{4.24}$$

By equation (4.13), we have

$$h = \frac{\bar{\epsilon}[x + z(1-q)]}{1 - q\bar{\epsilon}(z + \zeta)}. \tag{4.25}$$

If we multiply all the right hand sides of equations (4.24) with the denominator of h, i.e., the positive function $1 - q\bar{\epsilon}(z+\zeta)$, then we obtain a replicator equation with the same orbits. This is the replicator equation with $P_y = 0$,

$$P_x = 1 - q\bar{\epsilon}(z + \zeta), \tag{4.26}$$

$$P_\zeta = q\bar{\epsilon}[x + z(1-q)], \tag{4.27}$$

and

$$P_z = (1-q)[1 - q\bar{\epsilon}(z+\zeta)] + q\bar{\epsilon}[x + z(1-q)]. \tag{4.28}$$

We note that

$$P_z = P_\zeta + (1-q)P_x = (1-q)(1 - q\bar{\epsilon}\zeta) + q\bar{\epsilon}x. \tag{4.29}$$

For the average payoff $\bar{P} := xP_x + yP_y + zP_z + \zeta P_\zeta$, we obtain

$$\bar{P} = x + z(1-q). \tag{4.30}$$

Hence $P_\zeta = q\bar{\epsilon}\bar{P}$. Furthermore,

$$P_x - P_z = q[1 - \bar{\epsilon}(q\zeta + x + z)]. \tag{4.31}$$

There exists no fixed point in the interior of S_4, since $0 = P_y < P_\zeta < P_z < P_x$. The edges of S_4 are oriented as in figure 4.7: in particular, the edge $x = z = 0$ consists of fixed points. We note that for $\epsilon = 0$, the edge with $y = \zeta = 0$ also consists of fixed points. The function $V = (y/x)^{1-q}(z/\zeta)$ is an invariant of motion.

If we now consider the full replicator dynamics given by equations (4.21) through (4.23), (i.e., including the factor f), we see that $z + \zeta = c/bwq\bar{\epsilon}$ defines a plane consisting of fixed points. This is the set where f vanishes. The plane intersects the simplex if $q > c/wb\bar{\epsilon}$. In the prism with $z + \zeta > c/bwq\bar{\epsilon}$, the orientation of the orbits is preserved; in the complementary polyhedron, the orientation is reversed. There is a set of stable fixed points on the plane: this is where all four strategies stably coexist, see figure 4.8. If there are enough reciprocators, it pays to trust.

4.5 GROWING KNOWLEDGE

A population consisting only of reciprocators is unstable: it can be invaded by *AllC* players, and thus be replaced by an equilibrium of *AllC* players and reciprocators. This equilibrium is stable, but not asymptotically stable, as seen in figure 4.3. Indeed,

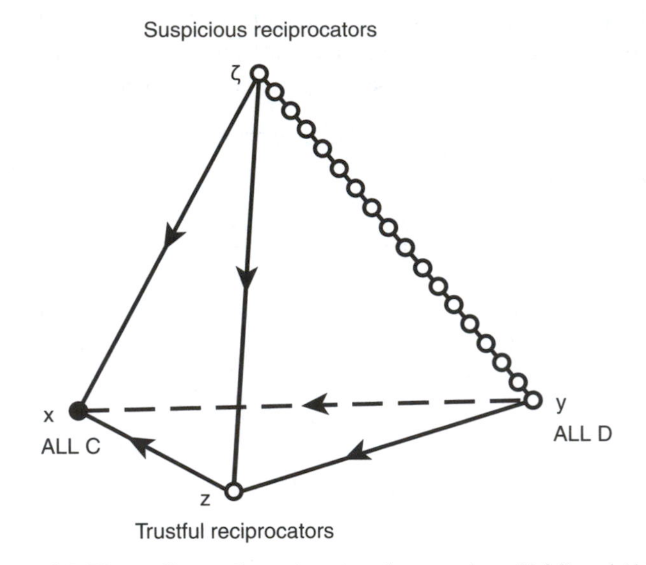

Figure 4.7 The replicator dynamics given by equations (4.24) and (4.25).

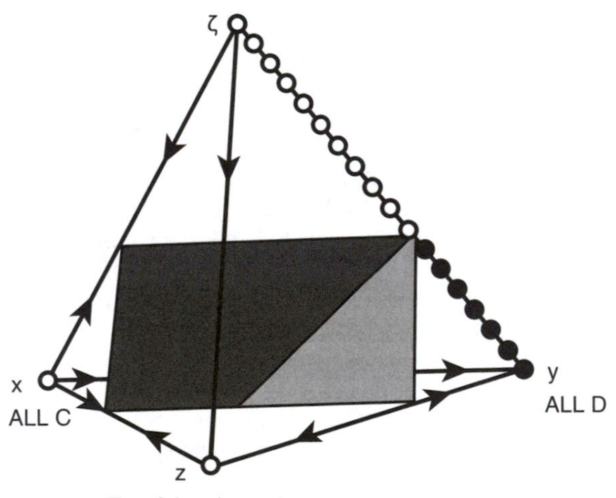

Figure 4.8 The replicator dynamics for *AllC*, *AllD*, trustful, and suspicious reciprocators, if the average information q is sufficiently large. The planar set consists of stable and unstable fixed points (the dark and the light area).

it is the end-point of a segment of rest points in S_3. It is easily conceivable that small random perturbations move the state of the population, along this segment, into a region where the rest points are no longer stable. From there, a small perturbation could lead back to the stable part of the line of rest points, but it could also lead to the basin of attraction of *AllD*, in which case cooperation breaks down.

It is instructive to compare this with the case of direct reciprocity, see figure 3.2. Here too, there exists a basin of attraction of *AllD*, and a sequence of random perturbations can, in the long run, lead any population state into this region, and thus destroy cooperation.

In the case of direct reciprocity, the simple model involving *AllC*, *AllD*, and *TFT* is just a first step, and further steps (for instance, the introduction of *Pavlov* or *Contrite TFT*) lead to much better prospects for cooperation. We shall see that the same holds in the case of indirect reciprocation. But before investigating other, more sophisticated strategies for indirect reciprocity, we analyze a modification of the initial model that turns the equilibrium of reciprocators and unconditional cooperators into a stable attractor.

Indeed, let us make the plausible assumption that the social network of each individual expands with time. This means that a player's probability of knowing a co-player's score is not constant but grows with the experience of the player. Let us assume that it is given by q_n in round n. Let w_n denote the probability that a randomly chosen individual is in round n (with $n=0$ as the initial round), and let q denote the average (taken over the whole population) of the q_n, i.e., $q := \sum w_n q_n$. If $q_n > q_{n-1}$ for all n, i.e., if players keep getting better and better informed about their co-players, we have, setting $q_{-1}=0$,

$$q > s := \sum w_n q_{n-1}. \tag{4.32}$$

It is easy to see that instead of equation (4.5) we now have

$$P_z(n) = -c\bar{\epsilon}(1 - q_n + q_n h) + b\bar{\epsilon}[x + z(1 - q + q\bar{\epsilon}(1 - q_{n-1} + q_{n-1}h))], \tag{4.33}$$

whereas the expressions for $P_x(n)$ and $P_y(n)$ remain unchanged, as in equations (4.3) and (4.4). It follows that

$$P_z(n) - P_x(n) = \bar{\epsilon}(1 - h)(cq_n - zb\bar{\epsilon}qq_{n-1}). \tag{4.34}$$

With $z_{cr} := c/b\bar{\epsilon}s$, we see that the total payoff values satisfy

$$P_x(z_{cr}) = P_z(z_{cr}). \tag{4.35}$$

Let us assume now that $z_{cr} < 1$, a condition slightly stronger than our previous condition $q > c/b\bar{\epsilon}$. Since equations (4.3) and (4.4) imply that for $n > 0$, $P_x(n) - P_y(n) = -c\bar{\epsilon} + b\bar{\epsilon}^2 qz$, we see that if $z = z_{cr}$, then

$$P_x(n) - P_y(n) = c\bar{\epsilon}(q - s)/s > 0 \tag{4.36}$$

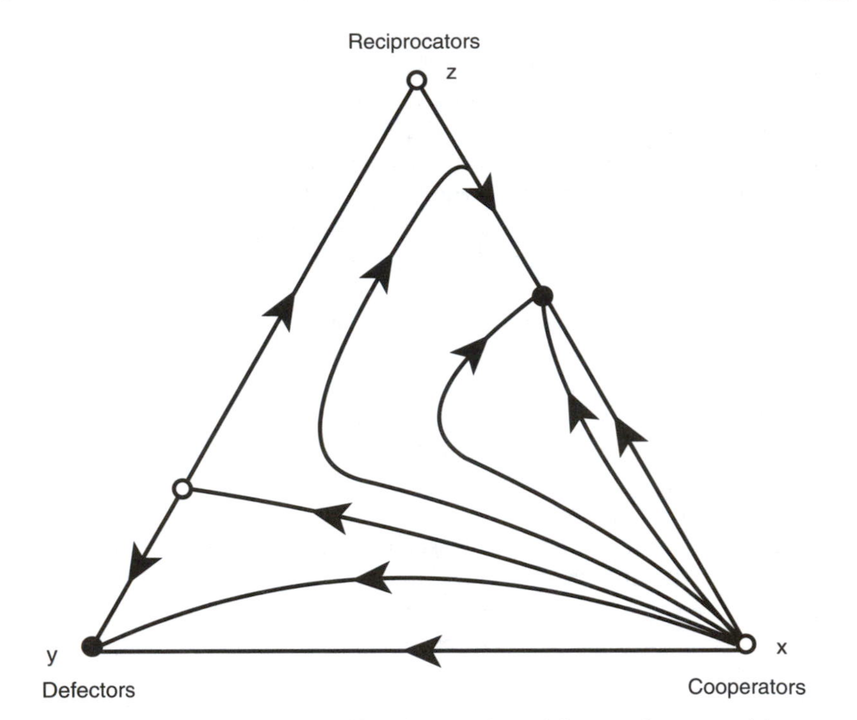

Figure 4.9 The replicator dynamics if each player's social network grows with age (and sufficiently many players experience more than one round).

holds for all $n > 0$. For the initial round, we have $P_x(0) - P_y(0) = -c\bar{\epsilon}$, which is negative. It follows that if w_0 is sufficiently small, i.e., if sufficiently many players experience several rounds, then

$$P_x(z_{cr}) > P_y(z_{cr}). \tag{4.37}$$

Hence there exists a mixture consisting of *AllC* players and reciprocators, $\mathbf{F}_{xz} = (1 - z_{cr}, 0, z_{cr})$, which cannot be invaded by defectors. The resulting replicator dynamics is bi-stable: one attractor consists of defectors only, the other is a mixture of reciprocators and unconditional altruists, see figure 4.9.

The previous result is somewhat spoiled by the fact that our standing assumption that w_n is proportional to w^n (see section 3.2) has to be dropped here. Indeed, in that case $(q - s)/s = (1 - w)/w$, and hence $P_x = P_y$ if $z = z_{cr}$. Too many players are experiencing one round only, and hence expression (4.37) does not hold. But many other choices for w_n work well, for instance the assumption that every player experiences N rounds.

4.6 JUSTIFIED REFUSAL

One aspect of the conditional strategy that we have considered so far seems paradoxical, and almost inconsistent. Indeed, why should a reciprocator ever refuse to donate? This affects the reciprocator's own image, and reduces the likelihood that this player will be helped by other reciprocators. Discriminating players will therefore be discriminated against. They are effectively policing the community, and may cause would-be defectors to change their mind, but they do so at a cost to themselves. How can such a strategy be selected?

One solution is almost obvious. It distinguishes between justified and non-justified defections. This distinction can be captured, for instance, in the notion of good and bad *standing*, which is similar to what we met in section 3.17 in the context of direct reciprocity. Players, accordingly, are born with a good standing, and keep it as long as they help players who are in good standing. These players can therefore keep their good standing even when they defect, as long as their defections are directed at players who are in bad standing.

However, "standing" is a rather complex notion, and seems to require a constant monitoring of the whole population, which may overtax the players. Indeed, suppose that your recipient A has, in a previous round, refused help a recipient B. Was this refusal justified? Certainly not if B had proved to be in good standing by giving help to all deserving recipients. But what if B had refused help to some player C? Then you would have to know whether B's defection towards C was justified. This means that you need to probe into the past of C, etc. With direct reciprocation, standing is much easier to handle. You have only to keep track of your co-player's previous interactions with yourself. Even here, an error in perception can lead to a deadlock: it may happen that both players believe that they are in good standing and keep punishing each other in good faith. With indirect reciprocation, the problem becomes much more severe: players have to keep track, not only of the antecedents of the current recipient, but also of the past actions of the recipient's former recipients, etc.

A priori, it is not obvious how individuals should update the images, or scores, or reputations, of their co-players. In fact, their standard of moral assessment, which eventually can lead to a social norm, can also be subject to evolution.

In the following sections on indirect reciprocation, we shall assume that individuals manage to keep track of the images of all players in their community (hence $q = 1$), and that they decide, when in the role of the donor, whether to give help or not, depending on their recipient's image, and possibly based on their own image. This can be viewed as an investigation of simple mechanisms for local information processing. But it has farther-ranging implications for the evolution of social norms, and hence of moral judgments. It concerns the issue of deciding when a defection is justified, and when not. In other words, when is a player good or bad?

Let us first consider this question in a very limited context, by assuming that the image, or score, can only take two values. Needless to say, one could also envisage many other strategies, taking into account, for instance, the accumulated payoffs for donor and recipient, or the prevalence of cooperation within the community, or whether one has received help in the previous round, etc.

Table 4.1 The Assessment Module

Situation/Strategy	Scoring	Standing	Judging
good → good	good	good	good
good → bad	good	good	bad
bad → good	good	good	good
bad → bad	good	good	bad
good ↛ good	bad	bad	bad
good ↛ bad,	bad	good	good
bad ↛ good	bad	bad	bad
bad ↛ bad	bad	good	bad

Note: This module specifies which image to assign to the potential donor of an observed interaction (good → bad means "a *good* player helps a *bad* player," bad ↛ good means "a *bad* player refuses to help a *good* player," etc).

4.7 BINARY MODELS: THE WORLD IN BLACK AND WHITE

We shall consider strategies specified by two modules, namely an assessment module and an action module. The *assessment module* of player C operates whenever C observes an interaction between two players A and B (see the diagram depicting vicarious reciprocity in fig. 4.1). In the eyes of the observer C, the image of the potential donor A may be affected by A's decision. The image of the potential recipient B, who is the passive party in the interaction, remains unchanged. The *action module* prescribes whether a player in the position of a potential donor actually provides help or not. This decision is based on the information obtained through that player's assessment module.

To start discussing the assessment module, we assume for simplicity that the score acquired by individual A in the eyes of individual C depends only on how A behaved when last observed by C in the role of a potential donor, i.e., on whether A gave or refused help to some third party B. Thus C has a very limited memory, and the score of A (in C's eyes) can take only two values, *good* or *bad*. In every interaction observed by C, there are two possible outcomes (A can give help or not), two possible score values for A, and two for B. Thus there are eight possible types of interaction, and hence, depending on whether they find C's approval or not, $2^8 = 256$ different assessment modules, or value systems.

As intuitively appealing examples of such assessment modules, let us consider in table 4.1 three of these value systems—embryonic moral systems, so to speak. We denote them as Scoring, Standing, and Judging, respectively. These assessment modules differ on which of the observed interactions incur reprobation and count as *bad*. Someone using the Scoring assessment rule will always frown upon any potential donor who refuses to help a potential recipient. This assessment rule was used by the reciprocator in the model discussed in sections 4.2 and 4.3: the assessment does not take into account the score of the players engaged in the interaction.

Table 4.2 The Action Module

Situation/Strategy	Self	Co	And	Or	AllC	AllD
good $\overset{?}{\rightarrow}$ good	no	yes	no	yes	yes	no
good $\overset{?}{\rightarrow}$ bad	no	no	no	no	yes	no
bad $\overset{?}{\rightarrow}$ good	yes	yes	yes	yes	yes	no
bad $\overset{?}{\rightarrow}$ bad	yes	no	no	yes	yes	no

Note: This module prescribes whether to help or not given the player's own image, and the image of the potential recipient (bad $\overset{?}{\rightarrow}$ good prescribes whether a player using this action module should give help when having a *bad* image and facing a *good* recipient, etc.).

By contrast, a player using the Standing assessment rule will condemn those who refuse to help a recipient having a *good* score, but will condone players who refuse to help a recipient having a *bad* score. Those using the Judging assessment system will, in addition, extend their reprobation to players who help a co-player having a *bad* score, and deem that *bad* players cannot improve their image by refusing to help *bad* recipients.

Thus these three value systems are of different strictness towards wrong-doers. Roughly speaking, someone who refuses to help is always bad in the eyes of a Scoring assessor. Only those who fail to give help to a *good* player are bad in the eyes of a Standing assessor. Someone who fails to give help to a *good* player, but also someone who gives help to a *bad* player is bad in the eyes of a Judging assessor (see Table 4.1).

We can classify assessment modules: they are said to be of *first order* if they depend only on the action taken, of *second order* if in addition they depend on the score of the recipient, and of *third order* if they depend, moreover, on the score of the donor. Scoring is of first order, Standing of second, and Judging of third order.

Turning to the action module, we assume that a potential donor's decision on whether to help or not is based entirely on the scores of the two players involved, namely the donor and the recipient. Since there are four situations (donor and recipient can each be *good* or *bad*), there are $2^4 = 16$ possible decision rules. Four intuitively appealing examples would be Co, Self, And, and Or (see Table 4.2). The Co player is uniquely affected by the score of the potential recipient, and gives if and only if that recipient's score is *good*. This is the action module of the reciprocator in the model considered so far. The Self players worry exclusively about their own score, and give help if and only if this score is *bad*. The And players give help if the recipient's score is *good* and their own score is *bad*, and the Or players give help if the recipient's score is *good* or their own score is *bad*. Of course the 16 decision rules also include the two unconditional rules, always to give, and never to give, *AllC* and *AllD*, which do not rely on scores at all.

A *strategy*, in this model for indirect reciprocity, is determined by a specific combination of an action and an assessment module. This yields altogether $2^4 \times$

Table 4.3 The Leading Eight Strategies

Situation/Strategy	L1	L2	L3	L4	L5	L6	L7	L8
good $\rightarrow$ good	good	good	good	good	good	good	good	good
good $\rightarrow$ bad	good	bad	good	good	bad	bad	good	bad
bad $\rightarrow$ good	good	good	good	good	good	good	good	good
bad $\rightarrow$ bad	good	good	good	bad	good	bad	bad	bad
good $\not\rightarrow$ good	bad	bad	bad	bad	bad	bad	bad	bad
good $\not\rightarrow$ bad	good	good	good	good	good	good	good	good
bad $\not\rightarrow$ good	bad	bad	bad	bad	bad	bad	bad	bad
bad $\not\rightarrow$ bad	bad	bad	good	good	good	good	bad	bad
good $\stackrel{?}{\rightarrow}$ good	yes	yes	yes	yes	yes	yes	yes	yes
good $\stackrel{?}{\rightarrow}$ bad	no	no	no	no	no	no	no	no
bad $\stackrel{?}{\rightarrow}$ good	yes	yes	yes	yes	yes	yes	yes	yes
bad $\stackrel{?}{\rightarrow}$ bad	yes	yes	no	no	no	no	no	no

Note: Each strategy is specified by an assessment module (the first 8 rows of the table) and an action module (the last 4 rows). These strategies obtain the highest payoff values, and are not invadable by defectors. Strategy L3 corresponds to Co-Standing, strategy L8 to Co-Judging. No Scoring strategy occurs in the list. The open assessment issues correspond to the situations good $\rightarrow$ bad , bad $\rightarrow$ bad, and bad $\not\rightarrow$ bad. Each of these eight assessment modules corresponds to a unique action module, which can be Or or Co.

$2^8 = 2^{12} = 4096$ strategies. They are not all different from each other: for instance, all strategies with an *AllD* action module are effectively equal, irrespective of their assessment module.

4.8 THE LEADING EIGHT

There nevertheless remains a plethora of strategies. Let us simplify their investigation by assuming, from now on, that each player's score is public knowledge. This implies that there exists only one assessment rule in the population. It also implies that $q = 1$: everybody knows the co-players' images. In that case, it turns out that only eight reciprocating strategies are *reasonable*, in the following sense:

(a) In a population consisting entirely of players using that strategy, a rare dissident using a different action module (but keeping the same assessment rule) cannot invade, provided the inequality $wb > c$ is satisfied.
(b) The average fitness per round in such a homogeneous population is equal to the theoretical maximum $b - c$ (or differs at most by a term of order ϵ, the probability of mis-implementing an intended donation).

The list of these strategies, named the *leading eight*, is given in Table 4.3.

Only the Co and the Or action modules occur among the leading eight. Such players always give help to a *good* player, and always defect, when *good*, against a *bad* player. The assessment modules of the leading eight are consistent with this prescription: they all assess players as *good* or *bad* if the players provide or refuse help to a *good* recipient, irrespective of their own score, and they all allow *good* players to refuse help to *bad* players without losing their reputation.

This leaves it open how to judge the action of a *good* player giving help to a *bad* player, of a *bad* player giving help to a *bad* player, and of a *bad* player refusing help to a *bad* player. These are precisely the 2^3 alternatives making up the leading eight. If the assessment module requires a *bad* player to give to a *bad* player, the corresponding action module is Or; in all other cases it is Co. We note that strategies with the Standing and the Judging assessment module can belong to the leading eight, but not those with the Scoring module. In particular, the simple reciprocating strategy considered in the first sections of this chapter, namely Co-Scoring, does not belong to the leading eight.

4.9 EXPLAINING THE LEADING EIGHT

Let us first neglect the possibility of errors. The two requirements (a) and (b) imply that within the homogeneous population of residents, players always cooperate, and cannot be invaded by *AllD* or *AllC* players. (It then will follow that no other action rule can invade either.)

Clearly, the success of a player is entirely defined by that player's actions towards the residents. Since an *AllD* player defects against a resident, and thus economizes the cost c, without obtaining a higher overall payoff (for otherwise, *AllD* could invade), this implies that the residents, who by requirement (a) always help other residents, must be compensated. This can only occur if they are more likely to receive help in the next round, i.e., if the defector receives no help (payoff 0). Since the next round occurs with probability w, the residents have an expected benefit wb. For compensating the residents, the condition $c < wb$ must therefore be satisfied.

Defectors can only be distinguished from the residents who helped other residents if they acquire another image; by convention, the defector will be labeled *bad* and the resident discriminator who helped another resident, *good*. This implies the two action rules $good \overset{?}{\to} good : yes$ and $good \overset{?}{\to} bad : no$ (rows 9 and 10 in table 4.3). Moreover, an action rule prescribing to help can only be advantageous if, by following it, one obtains a *good* image, while by not following it, one obtains a *bad* image. Hence row 9 implies rows 1 and 5 in table 4.3. Similarly, an action rule prescribing not to help can only be advantageous if by deviating from it, one cannot gain a better image. In particular, then, row 10 implies that either (i) $good \not\to bad$ is *good* (row 6), or (ii) both $good \not\to bad$ and $good \to bad$ are viewed as *bad*. But alternative (ii) implies that *bad* is infectious: *good* players meeting *bad* recipients become *bad*, no matter what they are doing. We shall presently see that we can discard this alternative.

Indeed, now let us admit the possibility of errors in implementing an intended donation. A player committing such a mistake will be branded as *bad*, and not helped in the next round by the residents. If being *bad* is infectious, the number of *bad* players cannot decrease; worse, since any round entails the probability of a mistake, this number will actually increase.

If there is no way to redress a *bad* image, a resident who defected erroneously will obtain no benefit in future rounds. If there are many rounds, this means that most players, eventually, will not benefit from any help. Hence there must be an action of a *bad* donor to redress the image. This must be an action performed towards a resident with a *good* image, since all other encounters are rare, and it can only be $bad \stackrel{?}{\rightarrow} good : yes$, since the other decision is indistinguishable from that of an *AllD* player. This in turn implies that $bad \rightarrow good$ is viewed as *good*, and $bad \not\rightarrow good$ is *bad*, i.e., row 11 implies rows 3 and 7.

We have now specified the action module in three positions (rows 9,10,11) and the assessment module in five positions (rows 1,3,5,6,7). The three rows 2, 4, and 8 of the assessment module remain open, and can be filled in 2^3 different ways. It is easy to see that for each of the eight resulting assessment modules, there exists exactly one action module yielding a strategy satisfying the requirements (a) and (b) from the previous section.

For instance, consider the L1 assessment module. Since $bad \rightarrow bad$ is *good*, and $bad \not\rightarrow bad$ is *bad*, it is clear that the optimal action rule requires that *bad* gives to *bad*, which yields the *Or* rule. The same holds for L2. Since *good* never gives to *bad*, the corresponding assessment rule is irrelevant. Conversely, *Or* is better than *Co* only for the assessment modules in the first two columns of table 4.3. Indeed, $bad \stackrel{?}{\rightarrow} bad : yes$ requires that it pays to help, i.e., that the image becomes *good* while by refusing to help, it stays *bad*. This does not hold for the other columns.

In order to better understand the common characteristics of the leading eight, we note that they satisfy the following properties:

- (A) Maintenance of cooperation. This means that *good* players cooperate with *good* players and that this is seen as *good* (rows 9 and 1).
- (B) Identification of defectors. If a *good* or a *bad* player refuses help to a *good* player, this is viewed as *bad* (rows 5 and 7).
- (C) Justified punishment. A *good* player meeting a *bad* player should refuse help without being branded as *bad* (rows 10 and 6). This excludes first order assessment modules, and in particular Scoring.
- (D) Apologies accepted. A mistaken defection should not lead to eternal damnation. Hence *bad* players can redeem themselves by cooperating with *good* players (rows 11 and 3).

It seems obvious that in a resident population of cooperating reciprocators, assessment rules and action rules should correspond. This requirement does not hold for Co-Scoring, as we have seen, since *good* players have to refrain from helping *bad* players although this makes them lose their *good* score. Interestingly, however, there are exceptions to this requirement among the leading eight: for the two strategies L7 and L8 of table 4.3, *bad* players meeting *bad* co-players cannot redress their score one way or the other. However, in a homogeneous population playing such a strategy, encounters between two *bad* players are exceedingly rare.

In a population consisting of only one of the leading eight and the unconditional strategy *AllC*, everyone cooperates. Both types of players are equally likely to defect by mistake, and hence to lose their *good* image. But against a player with a *bad* image, the reciprocators will defect, and hence spare themselves the cost c, without losing their *good* image. Therefore, they are better off, by a factor proportional to ϵ.

4.10 SECOND-ORDER ASSESSMENT

In second-order assessment, actions are judged according to whether help is given or not, and whether the recipient is *good* or *bad*, whereas the donor's score is not taken into account. This yields 16 different assessment modules. Two of them belong to the leading eight, namely L3 and L6 (see table 4.3). Strategy L3 has the Standing assessment rule (every action is *good* except defecting against a *good* recipient). According to the assessment module of L6, both defecting against a *good* recipient and helping a *bad* recipient are judged as *bad*, whereas everything else is *good*. Both strategies use Co as their action module.

Let us first consider the reciprocating strategy L6, together with *AllC*, and *AllD* (with frequencies z, x, and y, as usual) and perform the same analysis as in sections 4.2 and 4.3. We denote by h the frequency of players having a *good* image. These consist of (a) *AllC* players who have met a *good* player and committed no mistake, or who have met a *bad* player and committed a mistake (the probability for this is $h(1-2\epsilon)+\epsilon$); (b) *AllD* players who have met a *bad* player (probability $1-h$); and (c) reciprocators who have met a *good* player in the previous round and committed no mistake, or who have met a *bad* player (probability $1-h\epsilon$). Hence

$$h = \frac{1-\bar{\epsilon}x}{2y+z+\epsilon(2x+z)}. \tag{4.38}$$

For the payoffs in round $n \geq 1$, one obtains

$$P_x(n) = -c\bar{\epsilon} + b\bar{\epsilon}[x + z(h(1-2\epsilon)+\epsilon)], \tag{4.39}$$

$$P_y(n) = b\bar{\epsilon}[x + z(1-h)], \tag{4.40}$$

and

$$P_z(n) = -c\bar{\epsilon}h + b\bar{\epsilon}[x + z(1-h\epsilon)]. \tag{4.41}$$

A similar result holds for the payoff in the first round. Hence, up to the common factor $\bar{\epsilon}(1-w)$,

$$P_x - P_y = -c + wb\bar{\epsilon}z(2h-1), \tag{4.42}$$

$$P_z - P_y = -ch + wb\bar{\epsilon}hz, \tag{4.43}$$

and

$$P_z - P_x = (1-h)(c + wb\bar{\epsilon}z) > 0. \tag{4.44}$$

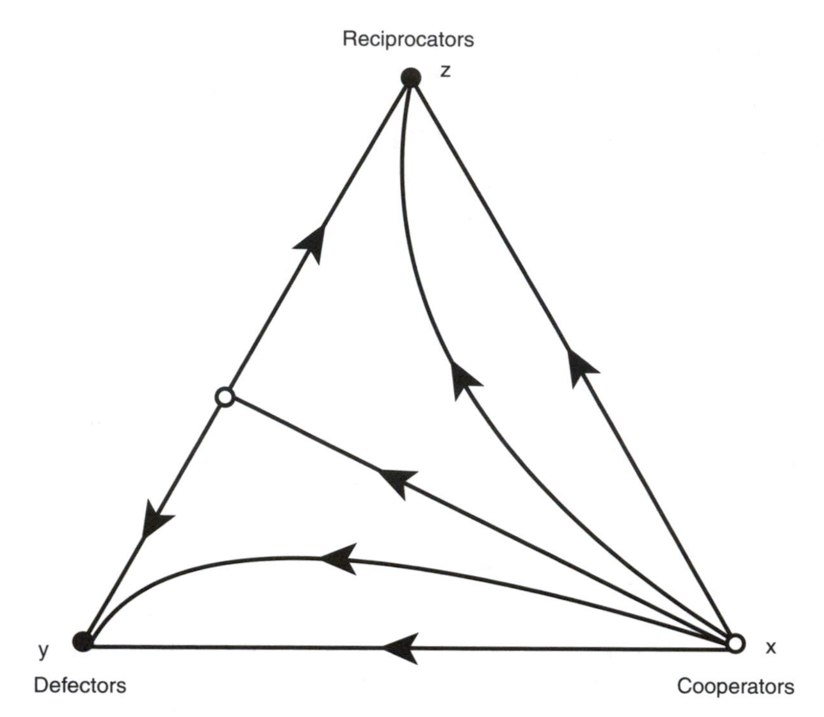

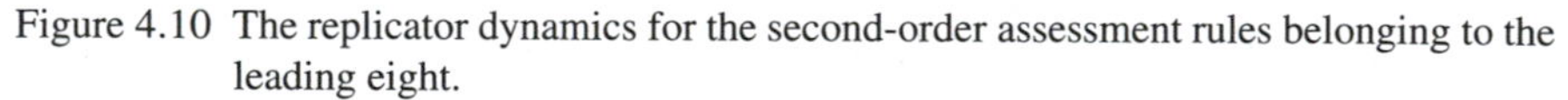
Figure 4.10 The replicator dynamics for the second-order assessment rules belonging to the leading eight.

Since the last expression is positive in $int S_3$, all orbits converge to the face $x = 0$. On this face, we have the usual bi-stability, with $\hat{z} = c/w\bar{\epsilon}b$ as unstable equilibrium, see figure 4.10.

Much the same holds for the second order strategy L3. *AllC* players have a good image if they have met a *good* player and committed no mistake, or if they have met a *bad* player: the probability is $1 - h\epsilon$. The same holds for the reciprocators. The *AllD* players are only *good* if they have met a *bad* player in the previous round, which has probability $1 - h$. This yields

$$h = \frac{1}{1 + y + \epsilon(x + z)}. \tag{4.45}$$

Clearly,

$$P_x(n) = -c\bar{\epsilon} + b\bar{\epsilon}[x + z(1 - h\epsilon)], \tag{4.46}$$

$$P_y(n) = b\bar{\epsilon}[x + z(1 - h)], \tag{4.47}$$

and

$$P_z(n) = -c\bar{\epsilon}h + b\bar{\epsilon}[x + z(1 - h\epsilon)]. \tag{4.48}$$

Again, $P_z - P_x = \bar{\epsilon}c(1 - h) > 0$ implies that $x \to 0$, and on that edge of the simplex S_3, there is an unstable equilibrium with $\hat{z} = c/bw\bar{\epsilon}$.

If mistakes in perception are included (in the sense that the majority knows the right reputation, but an ϵ minority gets it wrong), the attractor consisting of reciprocators is replaced by an attractor consisting of a mixture of reciprocators and unconditional cooperators. This assumption holds for the so-called "indirect observation model," which postulates that an interaction between A and B is observed by one player only, for instance C, and that all other members of the population adopt C's assessment. The situation becomes much more complex for a "direct observation model." In that model, all players keep their own private score of their co-players. Ultimately, it would seem that the evolution of assessment modules will have to be addressed in this context. It has been argued that thanks to language, all members of a population should agree on their scores, and that gossip is powerful enough to furnish all individuals with information about all past interactions. But it is common experience that even if two people witness the same interaction directly, they can differ in their assessment of that interaction. This suggests private scores.

4.11 REFERENCES

Indirect reciprocity was introduced by Sugden (1986) and Alexander (1987). The distinction between upstream and downstream reciprocity is due to Boyd and Richerson (1989). In classical game theory, repeated games against varying opponents have been studied by Rosenthal (1979), Kandori (1992), and Okuno-Fujiwara and Postlewaite (1995). The approach to indirect reciprocity via evolutionary games is due to Nowak and Sigmund (1998a, 1998b), see also Ferrière (1998). The important role of mistakes in stabilizing cooperation was pointed out in Lotem, Fishman, and Stone (1999). Ohtsuki (2004) investigated reactive strategies for indirect reciprocity. The continuous entry model from section 4.2 was introduced in Brandt and Sigmund (2005). For various means of turning the equilibrium between *AllC* and the reciprocating strategy into an attractor, we refer to Fishman (2003), Mohtashemi and Mui (2003), and Brandt and Sigmund (2004 and 2005). Diverse assessment rules allowing for justified defection have been proposed by Sugden (1986), Nowak and Sigmund (1998a), Leimar and Hammerstein (2001), Panchanathan and Boyd (2003), and Takahashi and Mashima (2004), see also Nowak and Sigmund (2005). The terminology is not standardized, and "standing" has different meanings with different authors. The classification of assessment and action modules was proposed by Brandt and Sigmund (2004) and Ohtsuki and Iwasa (2004). The leading eight were introduced by Ohtsuki and Iwasa (2004), and further studied in Ohtsuki and Iwasa (2006). The dynamics for second-order assessment was explored in Ohtsuki and Iwasa (2007). For more on misguided reciprocity (also denoted as generalized reciprocity), see Bshary and Grutter (2006), Engelmann and Fischbacher (2002), Pfeiffer et al. (2005), Nowak and Roch (2007), and Rutte and Taborsky (2007). Suzuki and Akiyama (2007a, 2007b) and Masuda and Ohtsuki (2007) investigated indirect reciprocity in larger groups. For more on binary norms, see Pacheco, Santos, and Chalub (2006), and Chalub, Santos, and Pacheco (2006). The effectiveness of reputation mechanisms is experimentally studied in Bolton, Katok, and Ockenfels (2004a, 2004b) and Keser (2003); the evolution of unconditional altruism through signalling benefits is in Lotem, Fishman, and Stone (2002). Milinski et al. (2001) present an experimental study of different assessment modules. Direct and indirect reciprocity are compared in Duwfenberg et al. (2001) and Roberts (2008), see also Pollock and Dugatkin (1992).

Chapter Five

Fairness and Trust: The Power of Incentives

5.1 ULTIMATE OFFERS

In the Ultimatum game (see section 1.11), two players are randomly assigned the role of Proposer and Responder. The experimenter then allocates a certain sum to the Proposer. The Proposer offers part of it to the Responder. If the Responder accepts, the sum is split accordingly, and the game is over. If the Responder declines, the whole sum returns to the experimenter, and again, the game is over: but now, both players receive nothing. It is important to stress that the two players know the rules in advance, and that they know that they will never meet again.

We shall normalize the sum to 1, and denote the size of the offer by p. The Proposer's strategy, then, is simply specified by $p \in [0, 1]$. The Responder's strategy is specified by the set of acceptable offers. It is plausible to assume that this set is an interval of the form $[q, 1]$. Hence the Responder's strategy is given by an aspiration level $q \in [0, 1]$. If $p \geq q$, then the Responder accepts the offer and obtains payoff p, whereas the Proposer's payoff is $1 - p$. If $p < q$, the offer is rejected and both players have payoff zero.

In most experiments, the Proposer offers between 40 and 50 percent, and this is accepted. The few offers below 20 percent are usually rejected. Most Proposers seem to anticipate this, and this prompts them to make a decent offer. But a selfishly motivated Responder should accept any positive offer, since it is better than nothing. Why then are aspiration levels usually well above 20 percent?

Proposer and Responder are in two distinct roles, which we denote by I and II. Each player's set of strategies is (in principle) the continuum of the unit interval $[0, 1]$. In practice, the set is finite, of course, due to the discrete nature of the monetary denomination. We shall simplify even further, and assume that the Proposer has only the choice between offering h or l, with $0 < l < h < 1$. For instance, the high offer h could be 40 percent and the low offer l, 15 percent of the total. The Responder could, in principle, accept both offers, one of them, or none. Again, we simplify by assuming that the Responder has to choose between two strategies only: the strategy that consists in accepting the high offer only, and the strategy that consists in accepting both possible offers.

Thus we are facing a reduced version. For role I, the two strategies $\mathbf{e}_1$ and $\mathbf{e}_2$ are given by the offers h and l; for role II, the two strategies $\mathbf{f}_1$ and $\mathbf{f}_2$ are again denoted by h and l, for convenience: these are now the Responder's aspiration levels. The payoff matrix is given by

$$
\begin{array}{c|cc}
 & \mathbf{f}_1 & \mathbf{f}_2 \\
\hline
\mathbf{e}_1 & (1-h, h) & (1-h, h) \\
\mathbf{e}_2 & (0, 0) & (1-l, l)
\end{array} . \tag{5.1}
$$

Two obvious Nash equilibrium pairs for this two-role game are (h, h) and (l, l).

Before analyzing the corresponding population dynamics, let us note that the Ultimatum game is, in fact, a symmetric game. It is only after a coin toss that the roles of Proposer and Responder are assigned to the two players. A strategy for the Ultimatum game is thus a pair $(p, q) \in [0, 1]^2$. For the reduced Ultimatum game, it is a pair $(\mathbf{e}_i, \mathbf{f}_j)$.

5.2 A MINI-COURSE ON MINI-GAMES

Before turning to the reduced Ultimatum in its symmetrized version, it is advisable to analyze the population dynamics of two-role games in greater generality. Let us consider a game with two roles I and II, with two strategies for each role, which we denote by $\mathbf{e}_i$ for role I and $\mathbf{f}_j$ for role II (with $i, j \in \{1, 2\}$). The payoff matrix is

$$
\begin{array}{c|cc}
 & \mathbf{f}_1 & \mathbf{f}_2 \\
\hline
\mathbf{e}_1 & (A, a) & (B, b) \\
\mathbf{e}_2 & (C, c) & (D, d)
\end{array} . \tag{5.2}
$$

As in section 2.5, we consider the symmetrized version (where a coin toss decides which role to assign to which player). The strategies for the resulting symmetric game will be denoted by $\mathbf{G}_1 = \mathbf{e}_1\mathbf{f}_1$, $\mathbf{G}_2 = \mathbf{e}_2\mathbf{f}_1$, $\mathbf{G}_3 = \mathbf{e}_2\mathbf{f}_2$ and $\mathbf{G}_4 = \mathbf{e}_1\mathbf{f}_2$. The payoff for a player using $\mathbf{G}_i$ against a player using $\mathbf{G}_j$ is given, up to the factor $1/2$ that we shall henceforth omit, by the (i, j) entry of the matrix

$$
M = \begin{pmatrix}
A+a & A+c & B+c & B+a \\
C+a & C+c & D+c & D+a \\
C+b & C+d & D+d & D+b \\
A+b & A+d & B+d & B+b
\end{pmatrix} . \tag{5.3}
$$

This corresponds to equation (2.14). For instance, a $\mathbf{G}_1$ player meeting a $\mathbf{G}_3$ opponent is in role I with probability $1/2$, plays $\mathbf{e}_1$ against the co-player's $\mathbf{f}_2$, and obtains B. The $\mathbf{G}_1$ player is in role II with probability $1/2$, plays $\mathbf{f}_1$ against the co-players' $\mathbf{e}_2$, and obtains c.

The replicator dynamics

$$
\dot{x}_i = x_i[(M\mathbf{x})_i - \mathbf{x} \cdot M\mathbf{x}] \tag{5.4}
$$

describes the evolution of the state $\mathbf{x} = (x_1, x_2, x_3, x_4) \in S_4$. Since the dynamics is unaffected if each m_{ij} is replaced by $m_{ij} - m_{1j}$ (for $i, j \in \{1, 2, 3, 4\}$), we can use the matrix

$$\begin{pmatrix} 0 & 0 & 0 & 0 \\ R & R & S & S \\ R+r & R+s & S+s & S+r \\ r & s & s & r \end{pmatrix}, \tag{5.5}$$

with $R := C - A$, $r := b - a$, $S := D - B$, and $s := d - c$. We shall denote this matrix again by M. It has the property that

$$m_{1j} + m_{3j} = m_{2j} + m_{4j} \tag{5.6}$$

for $j = 1, 2, 3, 4$. Hence

$$(M\mathbf{x})_1 + (M\mathbf{x})_3 = (M\mathbf{x})_2 + (M\mathbf{x})_4 \tag{5.7}$$

holds for all $\mathbf{x}$. From this and equation (2.32) it follows that the function $V = x_1x_3/x_2x_4$ satisfies

$$\dot{V} = V[(M\mathbf{x})_1 + (M\mathbf{x})_3 - (M\mathbf{x})_2 - (M\mathbf{x})_4] = 0 \tag{5.8}$$

in the interior of S_4, and hence that V is an invariant of motion for the replicator dynamics: its value remains unchanged along every orbit.

Therefore, the interior of the state simplex S_4 is foliated by the surfaces

$$W_K := \{\mathbf{x} \in S_4 : x_1x_3 = Kx_2x_4\}, \tag{5.9}$$

with $0 < K < \infty$. These are saddle-like surfaces that are spanned by the quadrangle of edges $\mathbf{G}_1\mathbf{G}_2$, $\mathbf{G}_2\mathbf{G}_3$, $\mathbf{G}_3\mathbf{G}_4$, and $\mathbf{G}_4\mathbf{G}_1$ joining the vertices of the simplex S_4, see figure 5.1.

The orientation of the flow on the edges can easily be obtained from the previous matrix. For instance, if $R = 0$, then the edge $\mathbf{G}_1\mathbf{G}_2$ consists of fixed points. If $R > 0$, the flow along the edge points from $\mathbf{G}_1$ towards $\mathbf{G}_2$, (which means that in the absence of the strategies $\mathbf{G}_3$ and $\mathbf{G}_4$, the strategy $\mathbf{G}_2$ dominates $\mathbf{G}_1$), and conversely, if $R < 0$, the flow points from $\mathbf{G}_2$ to $\mathbf{G}_1$.

Generically, the parameters R, S, r and s are non-zero. This corresponds to 16 orientations of the quadrangle $\mathbf{G}_1\mathbf{G}_2\mathbf{G}_3\mathbf{G}_4$, which by symmetry can be reduced to 4, see figure 5.2. Since $(M\mathbf{x})_1$ trivially vanishes, the fixed points in the interior of the simplex S_4 must satisfy $(M\mathbf{x})_i = 0$ for $i = 2, 3, 4$. This implies for $S \neq R$

$$x_1 + x_2 = \frac{S}{S - R}, \tag{5.10}$$

and for $s \neq r$

$$x_1 + x_4 = \frac{s}{s - r}. \tag{5.11}$$

Such solutions lie in the simplex if and only if $RS < 0$ and $rs < 0$, which corresponds to the orientations (c) and (d) of the quadrangle spanning the surfaces W_K. If this is the case, one obtains a line of fixed points that intersects each W_K in exactly one point (see fig. 5.1). The solutions can be written as

$$x_i = m_i + \xi \tag{5.12}$$

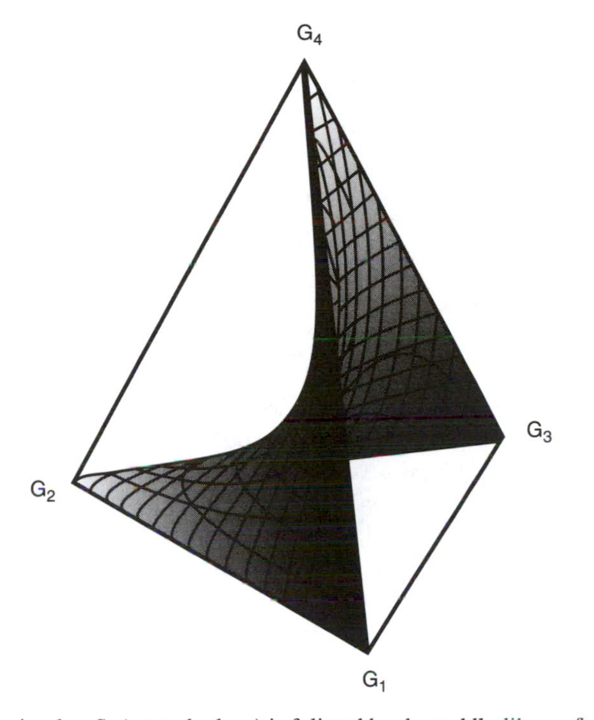

Figure 5.1 The simplex S_4 (a tetrahedron) is foliated by the saddle-like surfaces $W_K = \{\mathbf{x} \in S_4 : x_1x_3 = Kx_2x_4\}$, spanned by the edges $\mathbf{G}_1$–$\mathbf{G}_2$–$\mathbf{G}_3$–$\mathbf{G}_4$–$\mathbf{G}_1$. The surfaces are invariant for the replicator dynamics of the symmetrized game with two roles admitting two strategies each.

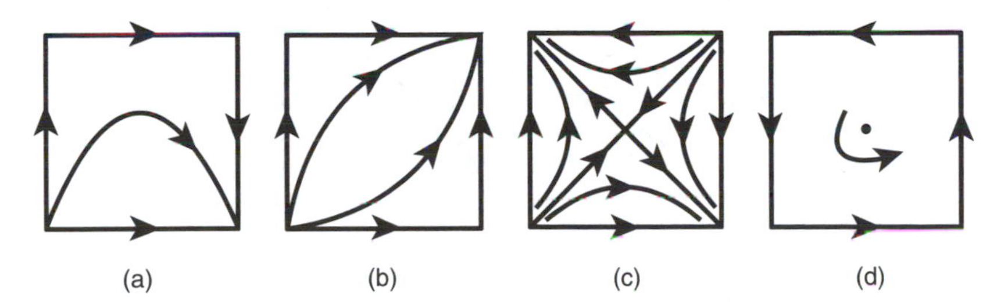

Figure 5.2 The four generic orientations of the cycle of the edges $\mathbf{G}_1$–$\mathbf{G}_2$–$\mathbf{G}_3$–$\mathbf{G}_4$ (up to symmetries). In cases (a) and (b), there is no fixed point in the interior of the square, and hence of the surfaces W_K. In cases (c) and (d), there exists such a fixed point.

for $i = 1, 3$, and

$$x_i = m_i - \xi \tag{5.13}$$

for $i = 2, 4$, with ξ as parameter and

$$\mathbf{m} = \frac{1}{(S-R)(s-r)}(Ss, -Sr, Rr, -Rs) \in W_1. \tag{5.14}$$

5.3 BACK TO THE REDUCED ULTIMATUM

Let us return to the reduced form of the Ultimatum game. The strategy $\mathbf{G}_1$ corresponds to (h, h): high offers, and a high aspiration level. We may view it as the *fair* or *pro-social* strategy. By contrast, $\mathbf{G}_3 = (l, l)$ epitomizes the *asocial* strategy. It enjoins acceptance of any positive offer, and parts with as little as possible. The strategy $\mathbf{G}_2 = (l, h)$ is *paradoxical*: it offers little, but insists on a high offer, in blatant contradiction to Kant's categorical imperative. Finally, $\mathbf{G}_4$ makes a good offer, but accepts a low offer. For want of a better term, we call it the *mild* strategy. The payoff parameters are $A = B = 1 - h$, $C = c = 0$, $D = 1 - l$, $a = b = h$, and $d = l$. Hence, $R = h - 1 < 0$, $r = 0$, $S = h - l > 0$, and $s = l > 0$. The asocial strategy dominates both the mild and the paradoxical strategy; the paradoxical strategy is dominated by both the pro-social and the asocial strategy; but the mild and the pro-social strategy are equivalent, in the absence of the other two strategies: indeed, all offers are fair. There exist no fixed points in the interior of S_4. Indeed, whenever $x_2 > 0$ or $x_3 > 0$, we have $(M\mathbf{x})_4 > (M\mathbf{x})_1$ and hence both ratios x_4/x_1 and x_3/x_2 are always increasing. On each surface W_K, the flow is as shown in figure 5.3.

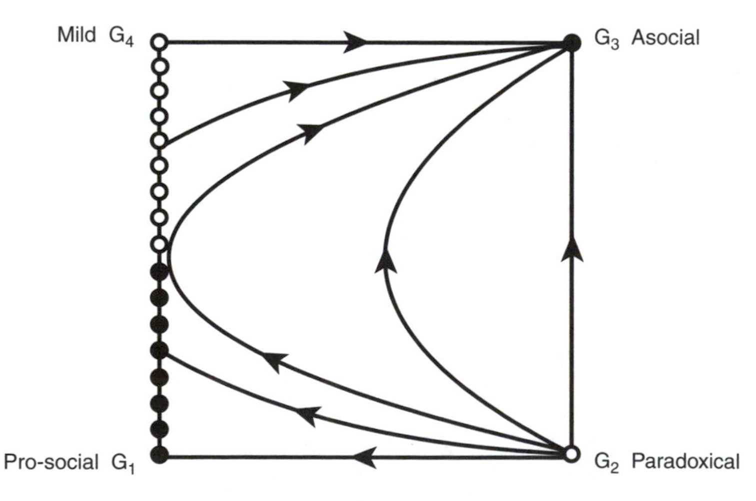

Figure 5.3 The replicator dynamics on surface W_K for the reduced Ultimatum game. The edge $\mathbf{G}_1\mathbf{G}_4$ consists of fixed points. In the long run, the evolution always leads to the asocial state $\mathbf{G}_3$.

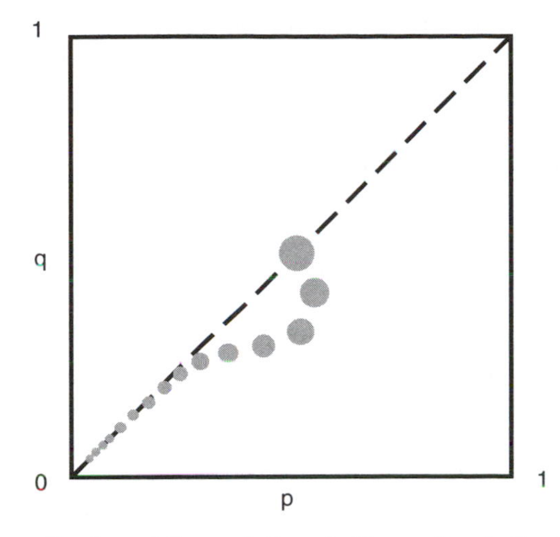

Figure 5.4 A schematic view of the evolution of offers and aspiration levels in individual-based simulations of the Ultimatum game, under conditions of strict anonymity. Both p and q converge to minimal values.

On the edge $x_2 = x_3 = 0$, all points are fixed points. If $x_1 < \frac{h-l}{1-l}$, then both $(M\mathbf{x})_2$ and $(M\mathbf{x})_3$ are larger than $\bar{M} = \mathbf{x} \cdot M\mathbf{x} = 0$. Let us denote by $\mathbf{Q}$ the point $(\frac{h-l}{1-l}, 0, 0, \frac{1-h}{1-l})$. Then the symmetric Nash equilibria of the game are the points on the segment $\mathbf{G}_1\mathbf{Q}$, and the vertex $\mathbf{G}_3$. We note that on the edge $x_2 = x_4 = 0$, there exists another fixed point $\mathbf{P}$, with coordinates $(h, 0, 1-h, 0)$. In a population with only pro-social and asocial players, we have a bi-stable competition. The pro-social strategy is risk-dominant (it has the larger basin of attraction) if $h < 1/2$.

The orbits in the interior of S_4 either converge to $\mathbf{G}_3$, or else to the segment of Nash equilibria. If we assume that random shocks occasionally perturb the state of the population, we will expect that they induce neutral drift along the edge $x_2 = x_3 = 0$. If $x_1 < \frac{h-l}{1-l}$, a random perturbation introducing the asocial strategy $\mathbf{G}_3$ will cause the fixation of $\mathbf{G}_3$. This implies that eventually, the population will consist of only asocial players. Thus evolutionary game theory leads to the same prediction as classical game theory; both are in contrast to experimental evidence.

It is interesting to return to the full Ultimatum game, with its continuum of strategies. We can perform an individual-based simulation, starting out with a population of 1000 individuals whose strategies (p, q) are randomly scattered over the strategy square $[0, 1]^2$. Let us assume that each individual plays 50 rounds of the Ultimatum game against 50 randomly chosen co-players. Then, players can update their strategy by imitating models chosen from the population with a probability proportional to their total payoff. Moreover, we assume that from time to time, an individual randomly choses a near-by strategy. This scenario based on imitation and innovation leads to a strategy very close to (0, 0) (see fig. 5.4), i.e., to an asocial population offering little and accepting anything—a far cry from human populations with their prevalent fairness norms.

5.4 BIFURCATION THROUGH REPUTATION

So far, we have considered conditions of strict anonymity. Let us now assume that with some (possibly small) probability, players may know their co-player by reputation, and in particular may know about the offers previously accepted by that co-player. Let us furthermore assume that occasionally, players offer less than they usually would, if they have reason to believe that they can get away with it, or more precisely, if they know that their co-player has previously accepted low offers. The two assumptions seem reasonable enough: they only require some information about other players in the group, and a touch of opportunistic self-interest. In that case, players accepting low offers face the risk that later offers proposed to them will tend to be low.

In order to analyze this situation, let us return again to the reduced Ultimatum, with two offers only, h and l. We assume that $\mu > 0$ is the probability that a fair (h, h) Proposer encountering a mild (h, l) Responder knows that this player is apt to accept a low offer, and consequently offers l instead of h. This yields the payoff matrix

$$\begin{array}{c|cc} & \mathbf{f}_1 & \mathbf{f}_2 \\ \hline \mathbf{e}_1 & (1-h, h) & (1-h+\mu(h-l), h-\mu(h-l)) \\ \mathbf{e}_2 & (0, 0) & (1-l, l) \end{array} \tag{5.15}$$

that differs from matrix (5.1) in the $(\mathbf{e}_1, \mathbf{f}_2)$ position only. The term $\mu(h-l)$, which may be arbitrarily small, can be viewed as a perturbation of the previous game.

The corresponding symmetrized game (5.5) is now given by $R = h - 1$, $r = -\mu(h-l)$, $S = (h-l)(1-\mu)$, and $s = l$. For $\mu < 1$, we have $R < 0$, $S > 0$, $s > 0$ (as before), and $r > 0$ (whereas we had $r = 0$ in the unperturbed case). This now yields a generic case, corresponding to case (c) in fig. 5.2. There exists a line of fixed points in the interior of the state space S_4. Each of the surfaces W_K (for $K > 0$) intersects this line in a saddle point. In particular, the fixed point on W_1 (where we have linkage equilibrium in the sense that $x_1x_3 = x_2x_4$, cf. section 3.6) is given according to equation (5.14) by

$$\mathbf{m} = \frac{1}{k}(l(h-l)(1-\mu), (h-l)^2\mu(1-\mu), (h-l)(1-h)\mu, l(1-h)) \tag{5.16}$$

with $k = (1-l-\mu(h-l))(l+\mu(h-l))$. For $\mu \to 0$, the point $\mathbf{m}$, and with it all interior fixed points, converge to the point $\mathbf{Q}$ on the edge $\mathbf{G}_1\mathbf{G}_4$.

The dynamics on each surface W_K is bi-stable, the vertices $\mathbf{e}_1$ and $\mathbf{e}_3$ are the attractors, see figure 5.5. Hence, depending on the initial condition, the population will either converge to the pro-social or to the asocial strategy. On the edge $x_2 = x_4 = 0$, the pro-social strategy $\mathbf{G}_1$ is risk-dominant if and only if $h + \mu(h-l) < 1/2$.

This shows that with reputation (and a small amount of selfishness), a population using the fair strategy cannot be invaded. The reason is that the "mild" strategy is now at a disadvantage. Again, we can perform individual-based simulations in the full Ultimatum game. This time we assume that a player's previous games can

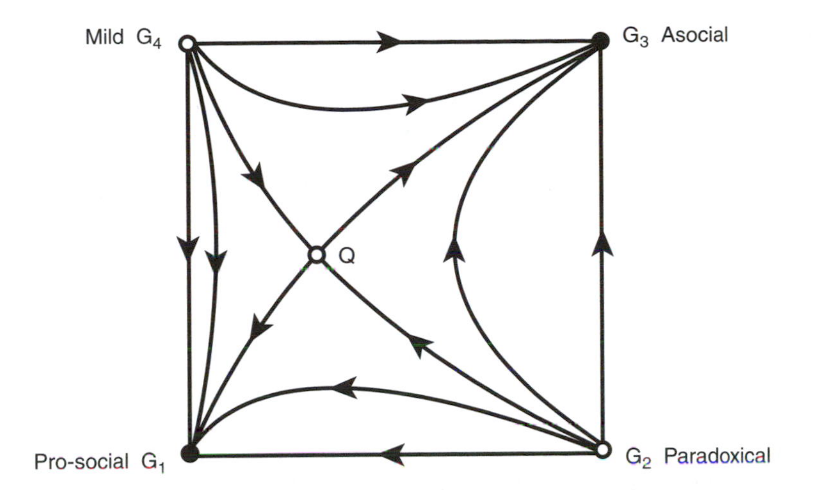

Figure 5.5 The replicator dynamics on the surfaces W_K for the reduced Ultimatum game, if Proposers know the offers accepted by their Responders in previous games. The dynamics is bi-stable, both the asocial and the pro-social state are attractors.

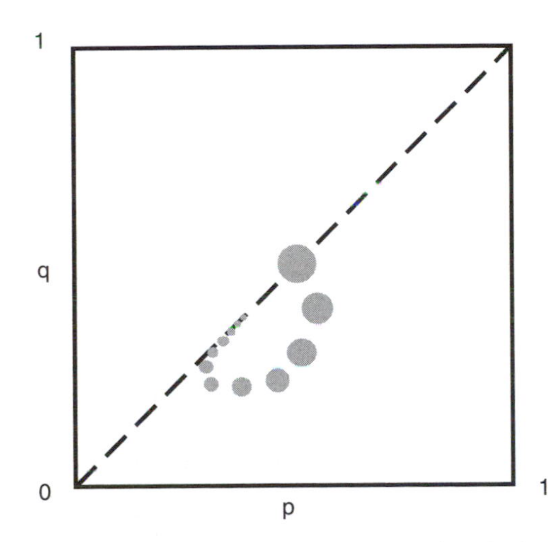

Figure 5.6 A schematic view of the evolution of offers and aspiration levels in individual-based simulations of the Ultimatum game, if players know each other's antecedents with a certain probability, and reduce their offer if they can get away with it. Both p and q converge to values slightly below 50 percent.

become known, and that Proposers offer whatever is smaller: either their p value, or the lowest amount known to have been accepted by the Responder during a previous game. The evolution leads first to very small (p, q) values, which then slowly creep up along the diagonal $p = q$. For a large range of parameter values, the offer tends toward some value between 40 and 50 percent, see figure 5.6.

5.5 DONATION AND DETERRENCE

Let us now return to the Donation game introduced in section 3.1. Two players must decide simultaneously whether or not to send a gift b to the other player, at a cost c to themselves. We know that the dominant solution is to defect, i.e., not to donate. But let us now add a second stage to this game. In this stage, players can harm their co-players. We shall assume that the harmful act consists of imposing a fine of size β. This fine is collected by the experimenter, not by the player imposing the fine. On the contrary, that player has to pay a fee, at a cost γ. The first stage of this game thus offers the possibility of altruism (helping another player at a cost to oneself), and the second stage the possibility of spite (harming the other player at a cost to oneself). Obviously, in both stages, the dominating solution is to avoid the cost. A self-regarding player should neither help nor harm the other player.

We assume that players impose their fine conditionally, fining only those who have failed to help them. This means that defectors can be *punished*. It is easy to see that the long-term outcome will still be the same as before: no pro-social behavior emerges. Indeed, let us label with $\mathbf{e}_1$ those players who cooperate by sending a gift to their co-player, and with $\mathbf{e}_2$ those who don't, i.e., who defect; similarly, let $\mathbf{f}_1$ denote those who punish defectors, and $\mathbf{f}_2$ those who don't. The payoff matrix is given by

$$\begin{array}{c|cc} & \mathbf{f}_1 & \mathbf{f}_2 \\ \hline \mathbf{e}_1 & (-c, b) & (-c, b) \\ \mathbf{e}_2 & (-\beta, -\gamma) & (0, 0) \end{array} \quad . \qquad (5.17)$$

We have used the same notation as for two-role games, although the situation is completely symmetric: instead of two roles, we now have two stages. Despite this difference, we can apply the same method as before. Indeed, each strategy for the two-stage game must specify what to do in the first stage, and what to do in the second. Hence, it is given by a pair $\mathbf{e}_i\mathbf{f}_j$ (with $i, j \in \{1, 2\}$). As in section 5.2, we denote the resulting four strategies with $\mathbf{G}_1 = \mathbf{e}_1\mathbf{f}_1$, $\mathbf{G}_2 = \mathbf{e}_2\mathbf{f}_1$, $\mathbf{G}_3 = \mathbf{e}_2\mathbf{f}_2$ and $\mathbf{G}_4 = \mathbf{e}_1\mathbf{f}_2$. The strategy $\mathbf{G}_1$ corresponds to the pro-social behavior: to give help, and to punish those who do not. $\mathbf{G}_3$ is the asocial strategy that avoids any costs: it neither helps the co-player, nor expects any help. $\mathbf{G}_2$ can again be viewed as paradoxical: a $\mathbf{G}_2$ player defects, but punishes co-players who defect. Finally, $\mathbf{G}_4$ can again be viewed as a mild strategy: a $\mathbf{G}_4$ player sends a gift to the co-player but does not react if nothing is returned.

We can follow the same approach as in section 5.2, and obtain a matrix of the form (5.5), with $R = c - \beta$, $S = c$, $r = 0$, and $s = \gamma$. Again, the manifolds $W_K = \{\mathbf{x} \in S_4 : x_1x_3 = Kx_2x_4\}$ are invariant (for $K > 0$) and the dynamics is as in figure 5.3. In fact, the reduced Ultimatum game is a special case, with $l = \gamma$, $\beta = 1 - l$, and $b = c = h - l$. Intuitively, this simply means that in the reduced Ultimatum game, the donation consists of making the high offer instead of the low offer. The benefit

$h - l$ to the Recipient (i.e., the Responder) is equal to the cost to the Donor (i.e., the Proposer). The punishment is in refusing the offer. This costs the Responder the amount l, and punishes the Proposer by the amount $1 - l$. This fine can be large if the offer has been dismal.

In the interior of S_4 (more precisely, whenever $x_2 > 0$ or $x_3 > 0$) we have $(M\mathbf{x})_4 > (M\mathbf{x})_1$ and hence x_4/x_1 is increasing. Similarly x_3/x_2 is increasing. Therefore there is no fixed point in the interior of S_4. Thus the fixed points in W_K are the vertices $\mathbf{G}_i$ and the points on the edge $\mathbf{G}_1\mathbf{G}_4$. $\mathbf{G}_3$ is a Nash equilibrium, $\mathbf{G}_2$ is not. On the edge $\mathbf{G}_1\mathbf{G}_4$, $\bar{M} = x_1(M\mathbf{x})_1 + (1 - x_1)(M\mathbf{x})_4$ vanishes (since $(M\mathbf{x})_4 = (M\mathbf{x})_1 = 0$). Hence a point $\mathbf{x}$ on that edge is saturated whenever $(M\mathbf{x})_3 \leq 0$, i.e., whenever $x_1 \geq c/\beta$. (The condition $(M\mathbf{x})_2 \leq \bar{M}$ reduces to the same inequality.) Thus if $c > \beta$, $\mathbf{G}_3$ is the only Nash equilibrium. This case is of little interest.

From now on, we restrict our attention to the case $c < \beta$: the fine costs more than the donation. We denote the point $(c/\beta, 0, 0, (\beta - c)/\beta)$ with $\mathbf{Q}$ and see that the closed segment $\mathbf{QG}_1$ consists of Nash equilibria.

On the edge $\mathbf{G}_2\mathbf{G}_4$, there exists a further fixed point $(0, c/(\beta + \gamma), 0, (\beta + \gamma - c)/(\beta + \gamma))$. It is attracting on the edge, and in the face $x_3 = 0$, but repelling on the face $x_1 = 0$. Finally, there is also a fixed point on the edge $\mathbf{G}_1\mathbf{G}_3$, namely the point $\mathbf{P} = ((c + \gamma)/(\beta + \gamma), 0, (\beta - c)/(\beta + \gamma), 0)$. It is attracting in the face $x_4 = 0$, but repelling in the face $x_2 = 0$. In the absence of the other strategies, the strategies $\mathbf{G}_1$ and $\mathbf{G}_3$ are bi-stable. The strategy $\mathbf{G}_1$ is risk-dominant (i.e., it has the larger basin of attraction) if, and only if $2c < \beta - \gamma$. We note that in the special case of the Ultimatum mini-game, this reduces to the condition $h < 1/2$.

Apart from $\mathbf{G}_3$ and the segment $\mathbf{QG}_1$ there are no other Nash equilibria. Depending on the initial condition, orbits in the interior of S_4 converge either to $\mathbf{G}_3$ or to a Nash equilibrium on $\mathbf{QG}_1$. In a population consisting of $\mathbf{G}_1$ and $\mathbf{G}_4$ only, no strategy has an advantage. We may assume that the state $\mathbf{x}$ fluctuates along that edge by neutral drift (reflecting random shocks of the system). Occasionally random shocks will also introduce a minority of the missing strategy $\mathbf{G}_2$ or $\mathbf{G}_3$. If this happens while $\mathbf{x}$ is in $\mathbf{QG}_1$, selection will send the state back to the edge, but a bit closer to $\mathbf{Q}$ (since x_4/x_1 increases). Once the state has reached the segment $\mathbf{QG}_4$ and a minority of $\mathbf{G}_3$ is introduced by chance, this minority will be favored by selection and eventually become fixed in the population. The asocial state $\mathbf{G}_3$ gets established in the long run.

5.6 DETERRENCE WORKS THROUGH REPUTATION

Let us assume that with a probability μ, cooperators (i.e., $\mathbf{e}_1$ players) defect against non-punishers, i.e., $\mathbf{f}_2$ players. (Hence μ is the probability that (1) the $\mathbf{f}_2$ type becomes known, and (2) the $\mathbf{e}_1$ type decides to defect.) Let us similarly assume that with a small probability ν, defectors (i.e., $\mathbf{e}_2$ players) cooperate against punishers. (Hence ν is the probability that (1) the $\mathbf{f}_1$ type becomes known, and (2) the $\mathbf{e}_2$ type decides to donate.) The payoff matrix becomes

$$\begin{array}{c|cc} & \mathbf{f}_1 & \mathbf{f}_2 \\ \hline \mathbf{e}_1 & (-c, b) & (-c(1-\mu), b(1-\mu)) \\ \mathbf{e}_2 & (-(1-\nu)\beta - \nu c, -(1-\nu)\gamma + \nu b) & (0, 0). \end{array} \quad (5.18)$$

We obtain $R = (1-\nu)(c-\beta) < 0$, $S = c(1-\mu) > 0$, $s = \gamma - \nu(b+\gamma)$, and $r = -b\mu < 0$. Thus the edge $\mathbf{G}_1\mathbf{G}_4$ no longer consists of fixed points, but of an orbit converging to $\mathbf{G}_1$. The dynamics is as in figure 5.5. The important parameter is μ (the probability that pro-social players defect if they know that they can get away with it). By contrast, as long as ν is small, it will not affect the dynamics. Therefore, we set $\nu = 0$ in this section.

There now exists a line of fixed points $\mathbf{x}$ in the interior of S_4, namely $x_i = m_i + \xi$ for $i = 1, 3$, and $x_i = m_i - \xi$ for $i = 2, 4$, with ξ as parameter and

$$\mathbf{m} = \frac{1}{(\gamma + b\mu)(\beta - c\mu)}(c\gamma(1-\mu), bc\mu(1-\mu), b\mu(\beta - c), \gamma(\beta - c)). \quad (5.19)$$

As with the reduced Ultimatum game, this line passes through the quadrangle $\mathbf{G}_1\mathbf{G}_2\mathbf{G}_3\mathbf{G}_4$ and hence intersects every surface W_K in exactly one point. Because $Rr > 0$, this point is a saddle point for the replicator dynamics on the corresponding surface W_K, see figure 5.5. On each surface, and therefore also in the interior of S_4, the dynamics is bi-stable, with attractors $\mathbf{G}_1$ and $\mathbf{G}_3$. Depending on the initial condition, every orbit, with the exception of a set of measure zero, converges to one of these two states.

Again, for $\mu \to 0$ the point $\mathbf{m}$, and consequently all interior fixed points, converge to the point $\mathbf{Q}$. At $\mu = 0$ we observe a highly degenerate bifurcation. The (very short) segment of fixed points is suddenly replaced by a transversal line of fixed points, namely the edge $\mathbf{G}_1\mathbf{G}_4$, of which one segment, namely $\mathbf{QG}_1$, consists of Nash equilibria.

Thus, introducing an arbitrarily small perturbation μ changes the long term behavior of the population. Instead of converging in the long run to the asocial regime $\mathbf{G}_3$ (defect, don't punish), the dynamics now has two attractors, namely $\mathbf{G}_3$ and the pro-social regime $\mathbf{G}_1$ (cooperate, punish defectors). We note that μ is proportional to the probability of having information about the co-player's type.

Let us briefly consider the case $\mu = 1$, which implies full knowledge about the type of the co-player. In this case, $S = 0$. This yields in some way the mirror image of the case $\mu = 0$. $\mathbf{G}_3\mathbf{G}_4$ is now the fixed point edge, the points on $\hat{\mathbf{Q}}\mathbf{G}_3$ are Nash, with $\hat{\mathbf{Q}} = (0, 0, b/(b+\gamma), \gamma/(b+\gamma))$, and fluctuations send the state ultimately to the unique other Nash equilibrium, namely $\mathbf{G}_1$. With perfect information, pro-social players gain the upper hand, if they are ready to defect whenever they can get away with it.

5.7 REVEALING ERRORS

The previous model is, in a certain sense, incomplete. It crucially relies on the fact that introducing reputation alters the dynamics on the edge $\mathbf{G}_1\mathbf{G}_4$. But on that edge, the population consists of two types only, both cooperating in the donation stage of the game. How should players learn whether the co-player is of type $\mathbf{f}_1$ or $\mathbf{f}_2$, i.e., willing to punish a defector, or not? Even if each player plays many rounds of the game, no defection ever arises.

There are several ways to deal with this question. One possibility consists of assuming that players learn about their co-players' propensity to punish from other sources. Indeed, it seems not unlikely that we can get a good idea about the irascibility or meekness of our co-players by watching their interactions with noisy children or their comments on the daily news, rather than merely from observing how they act in the Donation game.

But the simplest approach is to introduce errors. Let us assume that each player plays the game repeatedly (never against the same co-player twice, of course), and that players intending to donate will, with a certain probability ϵ, fail to implement their intention. (This could be due to a mistake, or to a momentary lack of resources.) In the absence of reputation, this yields the following payoff structure:

$$\begin{array}{c|cc} & \mathbf{f}_1 & \mathbf{f}_2 \\ \hline \mathbf{e}_1 & (-(1-\epsilon)c-\epsilon\beta,\,(1-\epsilon)b-\epsilon\gamma) & (-(1-\epsilon)c,\,(1-\epsilon)b) \\ \mathbf{e}_2 & (-\beta,\,-\gamma) & (0,0) \end{array} \quad (5.20)$$

Compared with the situation in section 5.5, s remains unchanged, whereas R and S are multiplied by $(1-\epsilon)$, which does not affect the sign, and hence conserves the dynamics on the corresponding edge. But r is now equal to $\epsilon\gamma$, and hence positive. This means that on the edge $\mathbf{G}_1\mathbf{G}_4$, the flow points towards $\mathbf{G}_4$: punishment is dominated. As a result, we obtain a dynamics as in case (a) of figure 5.2. All orbits in the interior of the simplex S_4 converge to the vertex $\mathbf{G}_3$. The asocial type wins.

Now let us introduce reputation. For simplicity, we will assume that a player who knows that the co-player is not of the punishing type never donates, i.e., always defects. (It would suffice to assume that the player defects with a small probability.) The parameter μ, then, is simply the probability to learn that the co-player has, on some past occasion, failed to punish a defection. If we assume perfect information, this occurs only if the co-player is of type $\mathbf{f}_2$ and has encountered a defection at least once. On the edge with $x_2 = x_3 = 0$, all players are willing to donate, and a defection only occurs by mistake. The probability that a co-player who experienced k rounds never faced a mistaken defection is $(1-\epsilon)^k$. If the number of rounds is distributed geometrically, with a constant probability $w < 1$ for a further round, then $w^k(1-w)$ is the probability that the co-player has experienced k rounds. This means that

$$\mu = \frac{w\epsilon}{1-w(1-\epsilon)}. \quad (5.21)$$

If we assume that an $\mathbf{e}_1$ player defects by mistake or when knowing that the co-player is of type $\mathbf{f}_2$, this yields

$$\begin{array}{c|cc} & \mathbf{f}_1 & \mathbf{f}_2 \\ \hline \mathbf{e}_1 & (-(1-\epsilon)c-\epsilon\beta,\,(1-\epsilon)b-\epsilon\gamma) & (-(1-\epsilon)(1-\mu)c,\,(1-\epsilon)(1-\mu)b) \\ \mathbf{e}_2 & (-\beta,\,-\gamma) & (0,0) \end{array} \quad (5.22)$$

We see that $r = \epsilon\gamma - \mu(1-\epsilon)b$ is negative if

$$\gamma < \frac{w(1-\epsilon)b}{1 - w(1-\epsilon)}, \tag{5.23}$$

i.e., if the fee for punishing the defector is not too high. In that case, the dynamics is as in figure 5.5, and the pro-social strategy $\mathbf{G}_1$ is an attractor.

Of course this can also be applied to the Ultimatum game. In that case, $r = \epsilon\gamma - \mu(1-\epsilon)b$ is negative if

$$l < w(1-\epsilon)h, \tag{5.24}$$

i.e., if the low offer is sufficiently smaller than the high offer.

5.8 THE TRUST GAME

The Trust game is a two-player game that forms an intriguing counterpart to the Ultimatum game. First, a coin toss decides who of the two players is the Proposer, or Investor. This Investor can then donate a certain sum c to the Responder, or Trustee, knowing that it will be multiplied by a factor $r > 1$. Next, the Trustee has the option to return some part β of this sum to the Investor (the returned amount will not be multiplied). This concludes the game.

It is obvious that a selfish Trustee ought to return nothing. Knowing this, the Investor should offer nothing. In real experiments, Investors often donate, and Trustees often return enough to make the exchange profitable to both.

In a reduced variant of the Trust game, we assume that the amounts c and β are fixed. The Investor has only to decide whether or not to send c to the Trustee. Thus an Investor has the choice between only two alternatives, namely $\mathbf{e}_1$ (donate) and $\mathbf{e}_2$ (defect). Similarly, a Trustee who receives a donation (i.e., the sum $b = rc$), has the choice between two alternatives, namely to return an amount β or not: these two alternatives will be denoted by $\mathbf{f}_1$ and $\mathbf{f}_2$. To make the game interesting, we will assume that $c < \beta < b$. In this case, if both players cooperate, they can both make a profit. The payoff matrix is

$$\begin{array}{c|cc} & \mathbf{f}_1 & \mathbf{f}_2 \\ \hline \mathbf{e}_1 & (\beta - c, b - \beta) & (-c, b) \\ \mathbf{e}_2 & (0, 0) & (0, 0) \end{array} . \tag{5.25}$$

Since the two players are in the role of Investor and Trustee with equal probability, they are effectively engaged in a symmetric game. Before analyzing it, we turn to another game that is only slightly more general, and exhibits an interesting complementarity to the Donation game with the option of punishing the defectors, studied in section 5.5. Indeed, let us consider a Donation game with the option of rewarding the donor. This is again a two-stage game. The first stage is simply the Donation game, as described in section 3.1. In the following stage, recipients have the option to return a part of their gift to the donor. We shall assume that this costs them γ, and yields β to the rewarded co-player (if $\beta = \gamma$, this is simply a payback). We assume

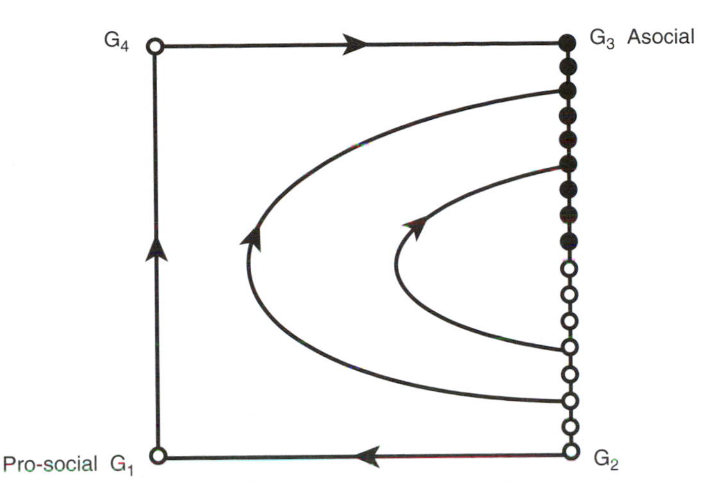

Figure 5.7 The replicator dynamics for the reduced Trust game. The edge $\mathbf{G}_2\mathbf{G}_3$ consists of fixed points. In the long run, nobody trusts the co-player.

$0 < c < \beta$ and $0 < \gamma < b$. If $\mathbf{e}_1$ and $\mathbf{e}_2$ are the two options for the first stage (to donate or to defect) and $\mathbf{f}_1$ and $\mathbf{f}_2$ for the second stage (to reward the donor or not), then the payoff structure is given by

	$\mathbf{f}_1$	$\mathbf{f}_2$
$\mathbf{e}_1$	$(\beta - c, b - \gamma)$	$(-c, b)$
$\mathbf{e}_2$	$(0, 0)$	$(0, 0)$

(5.26)

The reduced variant of the Trust game can be viewed as a special case of this (if we draw the usual parallel between two-role games and two-stage games). There exist four strategies, namely (a) the pro-social strategy $\mathbf{G}_1 = \mathbf{e}_1\mathbf{f}_1$ (donate, reward); (b) the distrustful strategy $\mathbf{G}_2 = \mathbf{e}_2\mathbf{f}_1$ (defect, but reward a donor); (c) the asocial strategy $\mathbf{G}_3 = \mathbf{e}_2\mathbf{f}_2$ (no donation, no reward); and finally, the strategy $\mathbf{G}_4 = \mathbf{e}_1\mathbf{f}_2$ (donation, but no reward). For the corresponding payoff matrix (5.5), we obtain $R = c - \beta < 0$, $r = \gamma > 0$, $S = c > 0$, and $s = 0$, see figure 5.7.

If $x_3 = x_4 = 0$, i.e., if everyone in the population is willing to reward a donation, then it is best to donate, i.e, $\mathbf{G}_1$ dominates $\mathbf{G}_2$. If $x_2 = x_3 = 0$, i.e., if donations can be taken for granted, then it is best not to reward, i.e., $\mathbf{G}_4$ dominates $\mathbf{G}_1$. If $x_1 = x_2 = 0$, i.e., if no one ever rewards a donation, then $\mathbf{G}_3$ dominates $\mathbf{G}_4$, i.e., it is best not to donate. Finally, if $x_1 = x_4 = 0$, so that nobody ever donates, then it does not matter whether one is prepared to reward a donation or not. In this case, every state of the population is a fixed point. Neither $\mathbf{G}_2$ nor $\mathbf{G}_3$ has an advantage.

It is easy to see that the segment $\mathbf{QG}_3$, with

$$\mathbf{Q} = \left(0, \frac{c}{\beta}, \frac{\beta - c}{\beta}, 0\right), \tag{5.27}$$

consists of saturated fixed points, i.e., of Nash equilibria. Indeed, for $x_1 = x_4 = 0$, both $(M\mathbf{x})_1$ (which is normalized to 0) and $(M\mathbf{x})_4$ are smaller than the average payoff $\bar{M} = (M\mathbf{x})_2 = (M\mathbf{x})_3 = c - \beta x_2$. In contrast to the situation in the reduced Ultimatum game, the flow along the edges leads from $\mathbf{G}_2$ to $\mathbf{G}_1$, from there to $\mathbf{G}_4$, and then to $\mathbf{G}_3$. All orbits in the interior converge to the segment $\mathbf{QG}_3$ for $t \to +\infty$ and to the segment $\mathbf{QG}_2$ for $t \to -\infty$. Thus the population will, in the long run, consist only of players who always defect (and consequently never reward). In particular, the reduced version of the Trust game will never get off the ground: no donations, no paybacks.

5.9 REWARD AND REPUTATION

Let us now introduce reputation effects. We shall assume that with some likelihood μ, cooperators (i.e., players of type $\mathbf{e}_1$) defect if they know that their co-player is not going to reward them (i.e., is of type $\mathbf{f}_2$). Hence μ is the probability that (1) the $\mathbf{f}_2$ type becomes known, and (2) the $\mathbf{e}_1$ type decides to defect. Similarly, we denote by ν the likelihood that defectors (i.e., players of type $\mathbf{e}_2$) cooperate if they know that they will be rewarded. Hence ν is the probability that (1) the $\mathbf{f}_1$ type becomes known, and (2) the $\mathbf{e}_2$-type reacts accordingly, and donates. This yields the payoff structure

	$\mathbf{f}_1$	$\mathbf{f}_2$
$\mathbf{e}_1$	$(\beta - c, b - \gamma)$	$(-c(1-\mu), b(1-\mu))$.
$\mathbf{e}_2$	$((\beta - c)\nu, (b - \gamma)\nu)$	$(0, 0)$

(5.28)

Now $R = (c - \beta)(1 - \nu) < 0$, $S = c(1 - \mu) > 0$, $r = \gamma - b\mu$, which is positive if μ is small, and $s = (\gamma - b)\nu$, which is negative. It is this last condition that differs from the unperturbed system studied in the previous section. The edge $\mathbf{G}_2\mathbf{G}_3$ no longer consists of fixed points. Instead, $\mathbf{G}_3$ is dominated by $\mathbf{G}_2$. The essential parameter, therefore, is ν, and we shall set $\mu = 0$ for most of the following discussion.

For $\nu > 0$, the flow on the edge $\mathbf{G}_2\mathbf{G}_3$ leads towards $\mathbf{G}_3$, so that the frame spanning the saddle-type surfaces W_K is cyclically oriented, see figure 5.8. As in section 5.6, there exists a line of fixed points in the interior of S_4. It can be shown that the surface W_1 consists of periodic orbits. If $\Delta := (\beta - \gamma)(1 - \nu) - (b - c)\nu$ is negative, all non-equilibrium orbits on W_K, with $0 < K < 1$, (as well as the orbits on the faces $x_1 = 0$ and $x_3 = 0$) spiral away from this line of fixed points and towards the heteroclinic cycle $\mathbf{G}_1\mathbf{G}_4\mathbf{G}_3\mathbf{G}_2$. All non-equilibrium orbits in W_K, with $K > 1$, (as well as the orbits on the faces $x_2 = 0$ and $x_4 = 0$) spiral away from that heteroclinic cycle and towards the line of fixed points. If Δ is positive, the converse holds.

We stress the highly unpredictable dynamics if $\nu > 0$ and $\Delta \neq 0$. The saddle-like surface W_1 divides the state space S_4 into two parts of equal size. For one-half of the initial conditions, the replicator dynamics sends the state towards the line of fixed points. But there, random fluctuations will eventually lead to the other half of the simplex, where the replicator dynamics leads towards the heteroclinic cycle $\mathbf{G}_1\mathbf{G}_4\mathbf{G}_3\mathbf{G}_2$. The population seems glued for a long time to one strategy, then suddenly switches to the next, remains there for a still longer time, etc. However, an

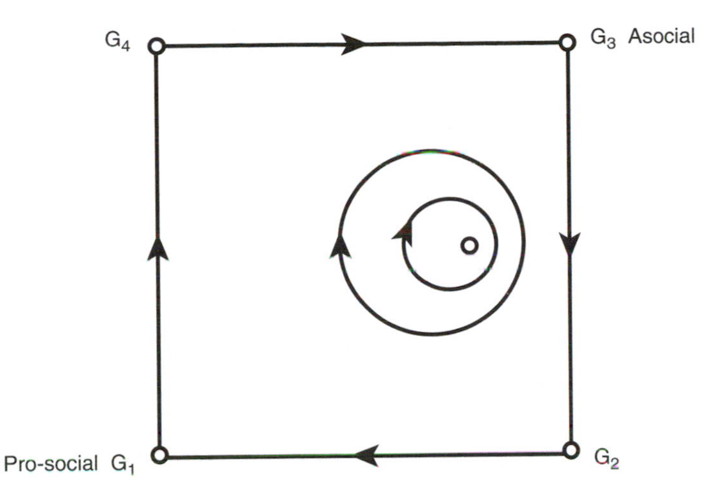

Figure 5.8 The replicator dynamics for the reduced Trust game, if receivers can gain a reputation for rewarding their donors. Depending on the initial value of the ratio x_1x_3/x_2x_4, the orbits either spiral towards the interior fixed point, or towards the cycle consisting of the four edges. If the ratio is 1, the orbits are closed.

arbitrarily small random shock can send the state back into the half-simplex where the dynamics drives it again towards the line of fixed points, etc. Not even the time averages of the frequencies of strategies converge. One can only say that the most probable state of the population is either monomorphic (i.e., close to one corner of S_4), or else close to the attracting part of the line of fixed points (with all four types present, and—if the value ν is small—a frequency of rewarders close to c/β, and a frequency of donations that is small).

Positive incentives thus appear to be considerably less efficient than negative incentives in furthering economic behavior. It needs to be stressed that this holds for models based on random pairing. If players can actively choose between partners, it seems likely that a reputation for rewarding helpful partners is more attractive than a reputation for punishing defectors.

Let us note that we encounter the same problem as for the Ultimatum game in section 5.7. If $x_1 = x_4 = 0$, then nobody ever donates. In this case, how should the trait of rewarding donations ever reveal itself? The assumption that occasionally players commit errors is not as plausible as in the previous case, since it is much less likely that an individual donates inadvertently than that a player fails in the intention to donate.

Finally, let us briefly consider what happens when the fact that a player does not reward is likely to become publicly known. In that case, such a player is not likely to receive a donation. This means that μ is close to 1, and hence that the parameter $r = \gamma - b\mu$ is negative. In that case, all orbits in the interior converge to $\mathbf{G}_1$, the pro-social state, see figure 5.9. This applies, in particular, to the Trust game played with banks and funds. As soon as it is safe to assume that a funds manager who fails to return the investment becomes known, the pro-social strategy (for the clients to

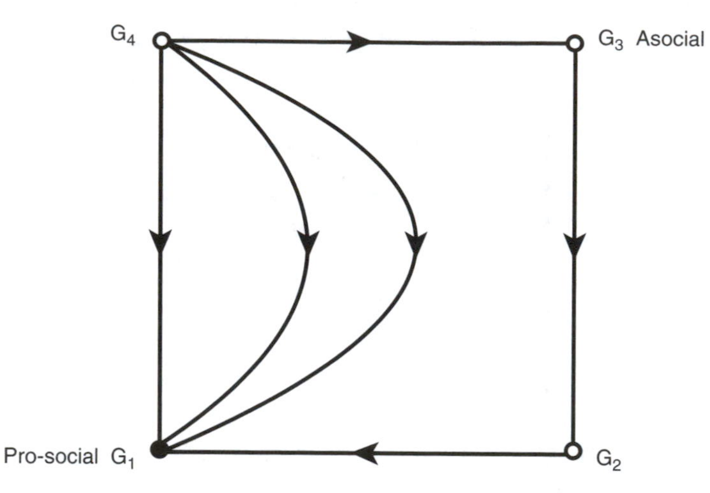

Figure 5.9 The replicator dynamics for the reduced Trust game, if players who do not return the investment are likely to become notorious. The pro-social state $\mathbf{G}_1$ attracts all orbits.

invest, and for the manager to return more than the invested sum to the clients) is a global attractor.

5.10 SNOWDRIFT ASYMMETRIES

In sections 2.5 and 5.2, we have seen how an asymmetric game can be transformed into a symmetric game. Occasionally, it can be useful to consider an opposite scenario, and allow asymmetries to creep into a symmetric game.

In particular, let us return to the Snowdrift game described in sections 1.4 and 3.1, but suppose that the two players are in distinct roles I and II. (For instance, one of them could be older than the other; or they could be of different sex.) The game is still a symmetric game (i.e., $\mathbf{e}_i = \mathbf{f}_i$ for $i = 1, 2$, and the payoff structure given by (5.2) satisfies $A = a$, $D = d$, $B = c$, and $C = b$), but the players can play conditional strategies of the type: if in role I, use $\mathbf{e}_1$, if in role II, use $\mathbf{f}_2$. Although the difference of the roles does not influence the payoff, it can be used as a cue. With the notation introduced in section 5.2, we see that $R = r$ and $S = s$. In the context of the Snowdrift game, $R = c/2 > 0$ and $S = c - b < 0$. The phase portrait on each manifold W_K is symmetric with respect to the diagonal, and bi-stable. The strategies G_2 and G_4 are attractors. They prescribe different moves in different roles. For almost all initial conditions, the state converges to one or the other of these two conditional strategies, which lead to conditional cooperation: players are ready to pay the fee or not, depending on the role they are in (see figure 5.10).

This means that an a priori irrelevant cue can settle the outcome. It could be age, for instance, or first arrival. The conditional strategy could be of the form: "If you arrived first on the spot, pay the fee; if you arrived second, do not." It could just as well be: "If you arrived second, pay the fee; if you arrived first, do not."

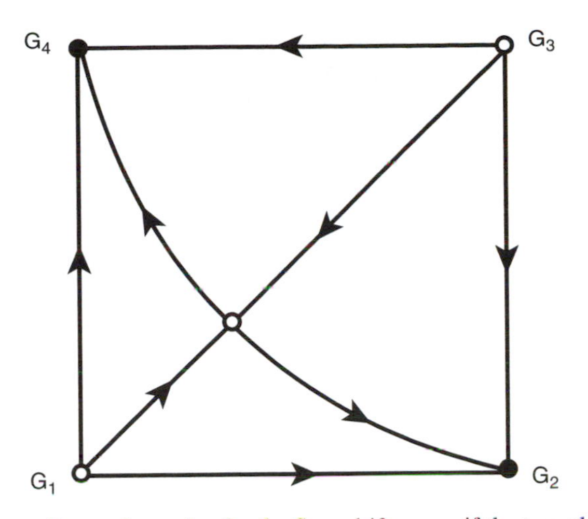

Figure 5.10 The replicator dynamics for the Snowdrift game, if the two players can be distinguished by some cue.

The symmetry between the two outcomes first has to be broken, one way or the other, through some exterior effect. Once the asymmetry is established, it is stable. A second asymmetry (for instance, in sex or size) will not supersede it.

For two-strategy games, this "desymmetrization" is only relevant in the case of stable co-existence. If the symmetric game leads to dominance or bi-stability, then the introduction of two roles does not affect the outcome: the corresponding asymmetric game still leads to dominance or bi-stability. But if the symmetric game leads to stable co-existence, then the introduction of two roles changes the outcome completely: the corresponding asymmetric game is bi-stable.

In particular, we have seen that in the repeated Snowdrift game, two players do equally well, on average, if they both cooperate in each round, than if they alternate in cooperating and defecting. But this second alternative appears unlikely. It seems more plausible that once a player has defected and the other has not, this asymmetry will be taken as a cue in the following rounds. In that case, one player will always cooperate and the other will always defect. Their joint payoff is still as good as if they fairly shared the cost of cooperation, round for round. But one player is consistently exploited by the other.

5.11 REFERENCES

The Ultimatum game was introduced by Güth, Schmittberger, and Schwarze (1982) and has played a huge role in experimental economy since, see e.g., Camerer (2003), Rankin (2003), Camerer and Fehr (2006), and Henrich (2006). For theoretical approaches in the context of evolutionary games, see Page, Nowak, and Sigmund (2000), Page and Nowak (2001), and Härdling (2007). Section 5.2 on symmetrized games follows Gaunersdorfer, Hofbauer, and Sigmund (1991), see also Hofbauer and Sigmund (1998). The reduced form of the Ultimatum game was analyzed in Nowak, Page, and Sigmund (2000). The individual-based simulations can be found in http://www.wu-wien.ac.at/usr/ma/hbrandt/ultimatum/. The effects of reward and punishment on the Donation game have been studied in Sigmund, Hauert, and

Nowak (2001), an extension to Public Good games can be found in Hauert, Haiden, and Sigmund (2004). The Trust game was introduced by Berg, Dickhaut, and McCabe (1995), see also Güth et al. (2001), Fehr and Fischbacher (2003), McCabe, Rigdon, and Smith (2003), Bohnet and Croson (2004), Cox (2004), and Eckel and Wilson (2004).Experiments by de Bruine (2005) show that trust is boosted by familiarity. Johnson and Bering (2006) argue that the fear of supernatural punishment plays an important role in the emergence of cooperation. Isaac and Walker (1988) and Rankin (2003) investigate the role of communication. Various counter-productive aspects of punishment are displayed in experiments by Gneezy and Rustichini (2000) or Fehr and Rockenbach (2003). The desymmetrization of symmetric games with stable co-existence has been studied (usually in the context of the escalation of conflicts) by Maynard Smith (1982), Sugden (1986), and many others.

Chapter Six

Public Goods and Joint Efforts: Between Freedom and Enforcement

6.1 PUBLIC GOODS GAMES

So far, we have considered pairwise interactions only. But many collaborative interactions take place in larger teams. This introduces new aspects. In particular, reciprocation becomes more difficult. If you interact repeatedly in a group, and if one of your co-players defects whereas another cooperates, with whom do you reciprocate?

On the other hand, in groups of more than two, majorities can form, and this may facilitate the enforcement of collaboration. As W. D. Hamilton wrote in his essay on *Innate Social Aptitudes of Man*: "There may be reason to be glad that human life is a many-person game and not just a disjoined collection of two-person games."

The team efforts that we shall consider will be modeled by so-called Public Goods games. Typical examples are group hunting or raiding, joint efforts in constructing shelters or preserving common resources, collaborations to ensure security from internal or external threats.

Such Public Goods games display a social dilemma: defectors do better than cooperators. The introduction of sanctions does not help to overcome that dilemma, because the sanctions are themselves a public good. But we shall see that if participation in the joint enterprise is voluntary, rather than compulsory, then cooperation based on the punishment of defectors can emerge.

6.2 MODELING PUBLIC GOODS GAMES

Let us assume that $N \geq 2$ individuals participate in a Public Goods game. Each has to decide whether to contribute towards the public good or not, i.e., whether to cooperate or to defect. We shall consider two models.

In the first case, which we denote by SR (for self-return), we assume that the contributions of all N_c cooperators are multiplied by $r > 1$ and then divided among all N players participating in the game. Thus every player receives the benefit

$$\frac{rcN_c}{N} \tag{6.1}$$

from the public good. In addition, the cooperators have to pay a fixed cost c. We note that a fraction r/N of their own investment returns to the contributors. The game can only qualify as a social dilemma if $r < N$.

In the second case, denoted by OO (for others-only), each contribution is multiplied by r and then divided among the $N-1$ *other* players. Thus contributors receive no return from their own investment. The payoff for a defector is

$$\frac{rcN_c}{N-1}. \tag{6.2}$$

A cooperator obtains

$$\frac{rc(N_c-1)}{N-1} \tag{6.3}$$

from the public good, and has moreover to pay a cost c.

If a player would switch from defection to cooperation, this would entail a cost, namely $c(1-\frac{r}{N})$ in the self-returning case, and c in the others-only case. If all players cooperate, they obtain $(r-1)c$ in both cases alike.

We note that if there are only two participants, i.e., if $N=2$, the OO Public Goods game yields the Donation game from section 3.1. The issue is simply whether or not to confer a benefit $b := rc$ to the co-player at a cost c to oneself. The SR Public Goods game also yields a Donation game, if $1 < r < 2$. In that case, the benefit is $rc/2$ and the cost is $c(1-\frac{r}{2})$.

Let us now consider the limiting case of an infinitely large population, and assume that from time to time, a random sample of N players engages in a Public Goods game. If x denotes the frequency of cooperators and y that of defectors, the expected payoff for a cooperator is given, in the OO case, by $P_x = c(rx-1)$, and that for a defector is given by $P_y = crx$. Multiplying these expressions by $(N-1)/N$ yields the corresponding payoff values in the SR case. In each case, P_y is larger than P_x, so that defectors will take over.

Can this social dilemma be overcome through positive or negative incentives specifically directed at individual players? We shall only investigate the effect of punishment here.

6.3 PUBLIC GOODS WITH PUNISHMENT

The Public Goods game with punishment has been studied in chapter 5 for the case of pairwise interactions, i.e., for $N=2$. It is easy to extend this analysis to larger N. The game consists of two stages. In the first stage, players have to choose between the alternatives $\mathbf{e}_1$ (to contribute to the public good), and $\mathbf{e}_2$ (to defect). In the second stage, players have to choose between $\mathbf{f}_1$ (to punish those who defected), and $\mathbf{f}_2$ (not to punish the defectors). As in section 5.5, we assume that each act of punishment reduces the payoff of the punished player by the amount β, and the payoff of the punishing player by an amount γ (with $\beta, \gamma > 0$). Thus, punishing is a costly activity: a selfish player should refrain from it. In the resulting two-stage game (first, contribute or not; then, punish or not), there are four distinct strategies. The pro-social strategy $\mathbf{G}_1 = \mathbf{e}_1\mathbf{f}_1$ contributes and punishes. The paradoxical strategy $\mathbf{G}_2 = \mathbf{e}_2\mathbf{f}_1$ prescribes to defect, and to punish all co-players who defect. The asocial strategy $\mathbf{G}_3 = \mathbf{e}_2\mathbf{f}_2$ consists in neither contributing nor punishing. Finally, the strategy $\mathbf{G}_4 = \mathbf{e}_1\mathbf{f}_2$ prescribes to contribute, but not to punish defectors. Players with this strategy can be viewed as *second-order exploiters*, free-riding on the sanctions provided by others.

We shall restrict attention in this section to the self-returning case SR. All players receive as a result of the contributions of their $(N-1)$ co-players an average payoff

$$B = \frac{rc}{N}(N-1)(x_1 + x_4). \tag{6.4}$$

The costs arising from their own contribution (if any) and the punishing activities in their group yield a net average payoff P_i for type $\mathbf{G}_i$, with

$$P_1 = B - c\left(1 - \frac{r}{N}\right) - (N-1)\gamma(x_2 + x_3), \tag{6.5}$$

$$P_2 = B - (N-1)\beta(x_1 + x_2) - (N-1)\gamma(x_2 + x_3), \tag{6.6}$$

$$P_3 = B - (N-1)\beta(x_1 + x_2), \tag{6.7}$$

$$P_4 = B - c\left(1 - \frac{r}{N}\right). \tag{6.8}$$

We consider the replicator equation $\dot{x}_i = x_i(P_i - \bar{P})$, where $\bar{P} = \sum x_i P_i$ is the average payoff. Since $P_1 + P_3 = P_2 + P_4$, the quotient x_1x_3/x_2x_4 denotes an invariant of motion and hence the sets

$$W_K = \{\mathbf{x} \in S_4 : x_1x_3 = Kx_2x_4\} \tag{6.9}$$

(with $K > 0$) provide a foliation of the state space S_4 into invariant manifolds, just as in figure 5.1. It is thus sufficient to study the dynamics on these two-dimensional manifolds, which are saddle-like surfaces spanned by the edges $\mathbf{G}_1 - \mathbf{G}_2 - \mathbf{G}_3 - \mathbf{G}_4 - \mathbf{G}_1$.

Just as in section 5.5, there is no rest point in the interior of these surfaces. The flow on the edge $\mathbf{G}_1\mathbf{G}_2$ points towards $\mathbf{G}_1$, and on the edges $\mathbf{G}_2\mathbf{G}_3$ as well as $\mathbf{G}_4\mathbf{G}_3$ it points towards $\mathbf{G}_3$. The edge $\mathbf{G}_1\mathbf{G}_4$ consists of fixed points. To make things interesting we shall always assume

$$(N-1)\beta > c\left(1 - \frac{r}{N}\right). \tag{6.10}$$

This condition states that the total fine imposed on a defector, if all co-players punish, is higher than the net cost of contributing to the public good. In this case, the point

$$\mathbf{Q} := \left(\frac{(N-r)c}{\beta N(N-1)}, 0, 0, 1 - \frac{(N-r)c}{\beta N(N-1)}\right) \tag{6.11}$$

lies in the segment $\mathbf{G}_1\mathbf{G}_4$. The growth rates $P_i - \bar{P}$ of the missing strategies 2 and 4 on the segment $\mathbf{G}_1\mathbf{Q}$ are negative, and hence these points are saturated, i.e., Nash equilibria. The points between **Q** and $\mathbf{G}_4$ are not. This yields a phase portrait as in figure 5.3. It follows that all initial conditions lead either to $\mathbf{G}_3$ or to the segment $\mathbf{G}_1\mathbf{Q}$. If we assume that random shocks occasionally perturb the system, we see that in the long run, the asocial equilibrium $\mathbf{G}_3$ always gets established.

6.4 REPUTATION

Let us now assume that with a small probability μ, (i) players occasionally learn about the type of their co-players, and (ii) players who normally are contributors may change their mind and decide not to contribute, if this entails no risk of being punished, i.e., if all co-players are of the non-punishing types $\mathbf{G}_3$ or $\mathbf{G}_4$. This alters the payoff values. Players with strategy $\mathbf{G}_i$ now have an expected payoff $P_i(\mu)$, with

$$P_1(\mu) = B - c\left(1 - \frac{r}{N}\right)[1 - \mu(x_3 + x_4)^{N-1}] - (N-1)\gamma(x_2 + x_3) \tag{6.12}$$

$$P_2(\mu) = B - (N-1)\beta(x_1 + x_2) - (N-1)\gamma(x_2 + x_3) \tag{6.13}$$

$$P_3(\mu) = B - (N-1)\frac{rc}{N}\mu(x_1 + x_4)(x_3 + x_4)^{N-2} - (N-1)\beta(x_1 + x_2) \tag{6.14}$$

$$\begin{aligned} P_4(\mu) = B &- (N-1)\frac{rc}{N}\mu(x_1 + x_4)(x_3 + x_4)^{N-2} \\ &- c\left(1 - \frac{r}{N}\right)[1 - \mu(x_3 + x_4)^{N-1}] \end{aligned} \tag{6.15}$$

where B remains unchanged, see equation (6.4). Indeed, the terms P_3 and P_4 for non-punishers are modified by the loss due to the change of mind of contributors: for each of the $N-1$ co-players, this happens if (a) the co-player is a contributor, and (b) all other $N-2$ co-players are non-punishers. The terms P_1 and P_4 for contributors are modified whenever all $N-1$ co-players are non-punishers.

Again, $P_1(\mu)+P_3(\mu) = P_2(\mu)+P_4(\mu)$ and hence the W_K are invariant manifolds. For small $\mu > 0$, the orientation of the flow on the edges $\mathbf{G}_1\mathbf{G}_2$, $\mathbf{G}_2\mathbf{G}_3$ and $\mathbf{G}_3\mathbf{G}_4$ remains unchanged, but the edge $\mathbf{G}_1\mathbf{G}_4$ no longer consists of rest points: the flow on this edge now points towards $\mathbf{G}_1$. The vertices $\mathbf{G}_1$ and $\mathbf{G}_3$ are sinks within each W_K, and $\mathbf{G}_2$ and $\mathbf{G}_4$ are sources, as can be seen by linearization. Thus there exists at least one rest point in the interior of each W_K. Moreover, there is only one such point, (which accordingly must be a saddle point, see fig. 5.5). Indeed, at the rest point, $P_1(\mu) = P_2(\mu)$ must hold. Setting $y = x_3 + x_4$ (the frequency of punishers) and

$$f(y) := P_1(\mu) - P_2(\mu), \tag{6.16}$$

we see that

$$f(y) = \mu\frac{(N-r)c}{N}y^{N-1} - \beta(N-1)y + \left[\beta(N-1) - \frac{(N-r)c}{N}\right]. \tag{6.17}$$

The equation $f(y) = 0$ has a unique solution $y = \hat{y}$ in $]0, 1[$ because f is strictly convex, $f(1) < 0$, and $f(0) > 0$. In addition, the rest point must satisfy $P_2(\mu) = P_3(\mu)$ and hence,

$$\gamma z = \frac{rc\mu}{N}(1 - z)\hat{y}^{N-2} \tag{6.18}$$

where $z = x_2 + x_3$ is the frequency of defectors. This specifies z. We note that for $\mu \to 0$ the interior rest point in W_K converges to $\mathbf{Q}$.

In the state space S_4, we therefore have a bi-stable situation: apart from a set of measure zero, all initial conditions lead to the pro-social or to the asocial state. Numerical simulations show that even for very small μ the basin of attraction of the social equilibrium $\mathbf{G}_1$ can be substantial.

The problem with this, and several other models is that they do not explain the emergence of social punishment. If all players are ready to punish, i.e., if the state is in the social equilibrium $\mathbf{G}_1$, then defectors obviously have no chance to invade. But in the asocial state $\mathbf{G}_3$, it is the pro-social trait that cannot gain a foothold. A player bent on punishing all defectors would have to punish left and right. This behavior would be very costly, and hence unlikely to be imitated.

6.5 FINITE POPULATIONS

In order to gain another perspective on the problem of the emergence of a sanctioning system, let us consider a finite population of size M. From time to time, a sample of N players is chosen at random and plays a Public Goods game. We consider three strategies only, which we denote by X, Y, and Z. The X players always contribute, but do not punish; the Y players neither contribute nor punish; and the Z players contribute and punish. In the previous model, this corresponds to $\mathbf{G}_4$ (the second-order exploiters), to $\mathbf{G}_3$ (the asocial players who neither contribute nor punish), and to $\mathbf{G}_1$ (the pro-social players, who contribute and punish the defectors). For simplicity, we do not include the paradoxical strategy $\mathbf{G}_2$ of defecting, but punishing defectors.

We shall now assume a particularly simple imitation mechanism. From time to time, two players are chosen at random and compare their payoffs. Whoever has the lower payoff adopts the strategy of the other player. If both players have the same payoff, a coin toss decides who imitates the other. If these updating events occur rarely enough, the payoff values, (which depend on the random sampling of the groups playing the Public Goods game), are very close to the expected values.

The corresponding stochastic process is given by a Markov chain whose absorbing states correspond to the homogeneous states $(M, 0, 0)$, $(0, M\ 0)$, and $(0, 0, M)$, which we denote by *AllX*, *AllY*, and *AllZ*. Imitation cannot lead away from these states. But we shall assume that in addition, players can occasionally switch to another strategy at random, without imitating another player. This “innovation” leads from a homogeneous state to a state with one dissident. Next, the imitation process takes over again. Either the dissident switches back to the resident strategy, or the rest of the population will eventually adopt the new strategy. The population then remains homogeneous again until the next random switch occurs, etc.

By assuming that these random switches are very rare, we effectively perform a separation of time scales: imitation works much faster than innovation. This "adiabatic" case has been described in section 2.17. The transitions from one homogeneous state to another can be described by a Markov chain with the three states *AllX*, *AllY*, and *AllZ*, given by

$$\begin{pmatrix} \frac{1}{2} - \frac{1}{2M} & \frac{1}{2} & \frac{1}{2M} \\ 0 & 1 & 0 \\ \frac{1}{2M} & 0 & 1 - \frac{1}{2M} \end{pmatrix}. \tag{6.19}$$

For instance, let us assume that the population is in state *AllX* and an individual switches randomly to another strategy. This is with probability $1/2$ the defector's strategy Y. Since Y players always do better than X players, the population will end up in *AllY*. With the same probability $1/2$, the random switch can introduce the pro-social strategy Z. In a population of X and Z players only, all do equally well, and the probability that eventually all individuals adopt the Z strategy (through neutral drift) is $1/M$. The transition probability from *AllX* to *AllX* is such that the row sum is 1. Now to the second row: in an *AllY* population, a single X individual will be exploited, and fare less well than the residents. Hence, nobody will imitate this individual, who will revert to Y on the next opportunity. Similarly, a single Z individual will do less well than the resident, provided that

$$c\left(1 - \frac{r}{N}\right) + \gamma(N-1) > \frac{N-1}{M-1}\left(\beta - \frac{rc}{N}\right), \tag{6.20}$$

in the self-returning case, and

$$c + \gamma(N-1) > \frac{N-1}{M-1}\left(\beta - \frac{cr}{N-1}\right) \tag{6.21}$$

in the others-only case. Both conditions are trivially satisfied if the total population size M is sufficiently large. Finally, an individual switching away from the *AllZ* state switches with equal probability to strategy X, or to strategy Y. In the former case, we have neutral drift again, and the fixation probability is $1/M$. In the latter case, the fixation probability is 0, provided we assume that a single defector does less well than the resident punishers. This is just condition

$$(N-1)\beta > \frac{N-1}{M-1}\gamma + c\left(1 - \frac{r}{N}\right)\left(\frac{M-N}{M-1}\right) \tag{6.22}$$

in the self-returning case, and

$$c\left(1 + \frac{r}{M-1}\right) < (N-1)\left(\beta - \frac{\gamma}{M-1}\right) \tag{6.23}$$

in the others-only case. In the limit of a large population, i.e., for $M \to \infty$, inequality (6.22) reduces to $(N-1)\beta > (1 - \frac{r}{N})c$ (as in equation (6.10)), and (6.23) to $(N-1)\beta > c$.

It is easy to see that the Markov chain given by matrix (6.19) has a unique stationary distribution, namely $(0, 1, 0)$. This shows that *AllY*, the pure defector state, is the inevitable outcome.

6.6 VOLUNTEERS STEPPING FORWARD

Now let us assume that there exists another strategy, denoted by W, which consists in not participating in the Public Goods game. Such players will obtain a payoff σ that does not depend on the other players. We shall assume that

$$0 < \sigma < (r-1)c. \tag{6.24}$$

This means that the payoff for a self-sufficient player who does not participate in the joint effort is lower than that obtained in a Public Goods game in which all members contribute, but higher than that in a Public Goods game among defectors only. We shall furthermore assume that single players cannot play a Public Goods game all by themselves. They need at least one more player willing to participate.

The Markov chain describing the transitions between the four homogeneous states *AllX*, *AllY*, *AllZ*, and *AllW* is given by

$$\begin{pmatrix} \frac{2}{3} - \frac{1}{3M} & \frac{1}{3} & \frac{1}{3M} & 0 \\ 0 & \frac{2}{3} & 0 & \frac{1}{3} \\ \frac{1}{3M} & 0 & 1 - \frac{1}{3M} & 0 \\ \frac{1}{6} & 0 & \frac{1}{6} & \frac{2}{3} \end{pmatrix}. \tag{6.25}$$

We only have to check the last row and column. (The other elements are obtained as in matrix (6.19), except that a random switch to another strategy now leads with equal probability $1/3$ to one of *three* alternatives.) It is clear that a single W player will do less well than the X residents, or the Z residents, but better than the Y residents. In a population of W players, a single individual switching to another strategy will do exactly as well as the residents, since that player cannot participate in any Public Goods game. Hence, if the lone dissident and one of the residents compare payoffs, the dissident is as likely to adopt the resident's strategy as vice versa. But if, as happens with probability $1/2$, a second individual adopts the dissident's strategy, then this strategy will fare less well than the W residents if it is the Y strategy, whereas it will fare better if it is the X or the Z strategy.

It is easy to see that the Markov chain (6.25) has a unique stationary distribution, given by

$$(p, p, 1 - 3p, p), \tag{6.26}$$

with

$$p = \frac{2}{M+8}. \tag{6.27}$$

For $M = 100$, this means that for almost 95 percent of the time, the population consists of the pro-social type only. Defectors are down to less than 2 percent, whereas they completely dominate the compulsory Public Goods game, as we have seen in the previous section.

6.7 THE OPTION TO ABSTAIN

In order to better understand the effect of voluntary participation, let us consider what happens if sanctioning is impossible, i.e., if only the strategies X, Y, and W are available, but not Z. The transition matrix then is

$$\begin{pmatrix} \frac{1}{2} & \frac{1}{2} & 0 \\ 0 & \frac{1}{2} & \frac{1}{2} \\ \frac{1}{4} & 0 & \frac{3}{4} \end{pmatrix}, \tag{6.28}$$

and the stationary distribution is

$$\left(\frac{1}{4}, \frac{1}{4}, \frac{1}{2}\right). \tag{6.29}$$

This means that in the long run, half of the time all the players participate in the game, either all contributing or all defecting, and half of the time none participate. The matrix (6.28) displays a Rock-Paper-Scissors structure: *AllX* can only mutate to *AllY*, which can only mutate to *AllW*, which can only mutate to *AllX*. It is this cycle that avoids the dead-lock of an all-defector state.

Let us analyze this cycle by turning to the replicator dynamics for an infinitely large population consisting of cooperators (who contribute, but do not punish), defectors, and non-participants. (Punishers are excluded.) We denote by x, y, and w the relative frequencies of the three strategies, and we calculate their expected payoff values P_x, P_y, and P_w. Clearly,

$$P_w = \sigma. \tag{6.30}$$

The probability that a given player has h co-players willing to participate is

$$\binom{N-1}{h}(1-w)^h w^{N-1-h} \tag{6.31}$$

for $h = 0, \ldots, N-1$. The probability that m of these are contributing is

$$\binom{h}{m}\left(\frac{x}{1-w}\right)^m \left(\frac{y}{1-w}\right)^{h-m} \tag{6.32}$$

for $m = 0, \ldots, h$ (if $h > 0$). If $h = 0$, the player is reduced to the non-participating strategy, and thus to the payoff σ.

Let us consider the OO case first. The expected benefit stemming from h co-participants is

$$\sum_{m=0}^{h} \frac{rcm}{h} \binom{h}{m} \left(\frac{x}{1-w}\right)^m \left(\frac{y}{1-w}\right)^{h-m} = \frac{rcx}{1-w}, \tag{6.33}$$

which is independent of h (for $h = 1, \ldots, N-1$). Hence the payoff obtained from the public good by a defector is

$$P_y = \sigma w^{N-1} + \frac{rcx(1-w^{N-1})}{1-w}. \tag{6.34}$$

Cooperators obtain the same term, reduced by $c(1-w^{N-1})$. Hence

$$P_y - P_x = c(1-w^{N-1}). \tag{6.35}$$

This expression is always positive.

In the SR case, when part of a contribution is returned to the donor, the expression is slightly more complicated. A defector in a group with h co-participants obtains $rcm/(h+1)$, if m is the number of cooperators. Hence the defector's expected payoff is

$$\left(\frac{rch}{h+1}\right)\left(\frac{x}{1-w}\right). \tag{6.36}$$

Thus,

$$P_y = \sigma w^{N-1} + rc\frac{x}{1-w} \sum_{h=0}^{N-1} \binom{N-1}{h} (1-w)^h w^{N-h-1} \left(\frac{h}{h+1}\right) \tag{6.37}$$

(the $h = 0$ term contributes nothing to the sum). The equality

$$\binom{N-1}{h} = \binom{N}{h+1} \frac{h+1}{N} \tag{6.38}$$

yields

$$P_y = \sigma w^{N-1} + rc\frac{x}{1-w}\left(1 - \frac{1-w^N}{N(1-w)}\right). \tag{6.39}$$

In a group with h co-players participating in the self-returning Public Goods game, switching from cooperation to defection yields $c(1 - r/(h+1))$. Hence,

$$P_y - P_x = \sum_{h=1}^{N-1} c\left(1 - \frac{r}{h+1}\right)\binom{N-1}{h}(1-w)^h w^{N-h-1}. \tag{6.40}$$

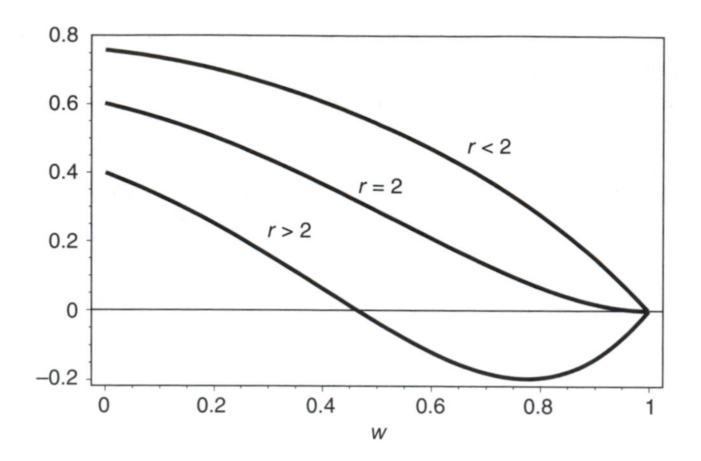

Figure 6.1 The graph of the function $w \mapsto (1 - w)F_N(w)$, with F_N given by expression (6.42), for $N = 5$.

Using equation (6.38) as before, we obtain

$$P_y - P_x = cF_N(w), \tag{6.41}$$

with

$$F_N(w) := 1 + (r-1)w^{N-1} - \left(\frac{r}{N}\right)\left(\frac{1-w^N}{1-w}\right). \tag{6.42}$$

The advantage of defectors over cooperators depends only on the fraction of non-participants w.

The sign of $P_y - P_x$, determines whether it pays to switch from cooperation to defection or not, $F_N(w) = 0$ being the equilibrium condition. We claim that for $r \leq 2$, F_N has no root, and for $r > 2$ exactly one root $\hat{w}$ in the interval $]0, 1[$. In order to show this, we consider the function $G(w) = F_N(w)(1 - w)$ that has the same roots as $F_N(w)$ in $]0, 1[$, and note that $G(0) = 1 - r/N > 0$ and $G(1) = 0$. For $r > 2$, the function G has a local maximum at $w = 1$. Indeed, $G'(1) = 0$ and $G''(1) = (2 - r)(N - 1)$. Moreover, $G''(w) = w^{N-3}(N-1)[(N-2)(r-1) - w(Nr - N - r)]$ changes sign at most once in $]0, 1[$. Thus, for $r > 2$ there exists a threshold value $\hat{w}$ of the frequency of non-participants above which cooperators fare better than defectors, see figure 6.1.

The average payoff in the population, $\bar{P} = xP_x + yP_y + wP_w$, can be rewritten using $y = 1 - x - w$:

$$\bar{P} = x(P_x - P_y) + w(\sigma - P_y) + P_y = -x(P_y - P_x) + (1-w)(P_y - \sigma) + \sigma. \tag{6.43}$$

This yields

$$\bar{P} = \sigma + [(r-1)xc - (1-w)\sigma](1 - w^{N-1}) \tag{6.44}$$

both for the self-returning and the others-only case.

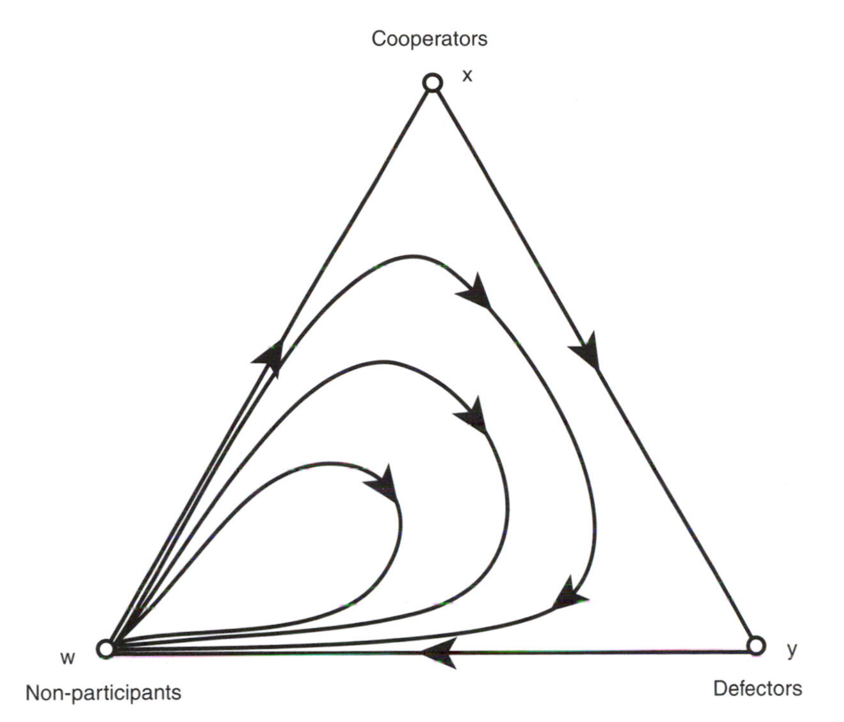

Figure 6.2 The replicator dynamics for the OO case, or for the SR case with $1 < r \leq 2$. All orbits in the interior of S_3 tend to $w = 1$, for $t \to \pm\infty$.

6.8 A ROCK-PAPER-SCISSOR DYNAMICS

Let us now analyze the replicator dynamics for the three strategies X, Y, and W, i.e., the cooperators, the defectors, and the non-participants. The behavior on the boundary of S_3 is the same for the SR and the OO scenario. The corners of the simplex $S_3 = \{(x, y, w) : x, y, w \geq 0, x + y + w = 1\}$, i.e., the homogeneous states *AllX*, *AllY*, and *AllW*, are obviously fixed points. There are no other fixed points on the boundary of S_3. In fact, an orbit leads from *AllX* (cooperators only) to *AllY* (defectors only), an orbit from *AllY* to *AllW* (non-participants), and an orbit from *AllW* to *AllX*. Thus the boundary of S_3 consists of a heteroclinic cycle.

In the OO case, we always have $P_x < P_y$. In the interior of the state space S_3, there is no fixed point: all orbits converge to *AllW*, for $t \to \pm\infty$ (see fig. 6.2).

For the rest of this section we only consider the SR case. For $r \leq 2$, the function F_N defined in expression (6.42) has no root in $]0, 1[$, and hence $P_x < P_y$ in $int S_3$: this leads to the same dynamics as in the OO case, see figure 6.2. Thus let us now assume $r > 2$.

In that case, F_N has a root $\hat{w} \in]0, 1[$. Using $P_y = P_w = \sigma$, we see that there exists a unique rest point $\mathbf{Q} = (\hat{x}, \hat{y}, \hat{w})$ in $int S_3$, with

$$\hat{x} = \frac{\sigma}{c(r-1)}(1 - \hat{w}), \tag{6.45}$$

as well as

$$\hat{y} = \left(1 - \frac{\sigma}{c(r-1)}\right)(1 - \hat{w}). \tag{6.46}$$

In order to analyze the dynamics in the interior of S_3, it is useful to show that the replicator equation can be rewritten in the form of a Hamiltonian system, and thus admits an invariant of motion. Indeed, defining $f = x/(x+y)$ as a new variable, we obtain

$$\dot{f} = \frac{xy}{(x+y)^2}(P_x - P_y). \tag{6.47}$$

This yields

$$\dot{f} = -f(1-f)cF_N(w). \tag{6.48}$$

Using equation (6.44), we see that

$$\dot{w} = [\sigma - cf(r-1)]\, w(1-w)(1-w^{N-1}). \tag{6.49}$$

Dividing the right hand sides of equations (6.47) and (6.48) by the function $f(1-f)w(1-w)(1-w^{N-1})$, which is positive for all values of f and w between 0 and 1, corresponds to a change in velocity and does not affect the orbits. This yields

$$\dot{f} = \frac{-cF_N(w)}{w(1-w)(1-w^{N-1})} =: -g(w) \tag{6.50}$$

$$\dot{w} = \frac{\sigma - cf(r-1)}{f(1-f)} =: l(f). \tag{6.51}$$

Introducing $H := G + L$, where $G(w)$ and $L(f)$ are primitives of $g(w)$ and $l(f)$, we thus obtain the Hamiltonian system

$$\dot{f} = -\frac{\partial H}{\partial w}, \tag{6.52}$$

$$\dot{w} = \frac{\partial H}{\partial f}. \tag{6.53}$$

This system is conservative, and the Hamiltonian H attains a strict global maximum at $(\frac{\sigma}{c(r-1)}, \hat{w})$. Thus the change in variables from $(x, y, w) \in S_3$ to $(f, w) \in]0, 1[^2$ shows that the unique interior equilibrium $\mathbf{Q}$ of the replicator dynamics is a stable point surrounded by closed orbits, see figure 6.3.

Variations of the three parameters N, r, and σ allow for $\mathbf{Q}$ to be in any interior point of the simplex (see fig. 6.4). The fixed point $\mathbf{Q}$ lies on the line

$$x = \frac{\sigma}{c(r-1) - \sigma} y, \tag{6.54}$$

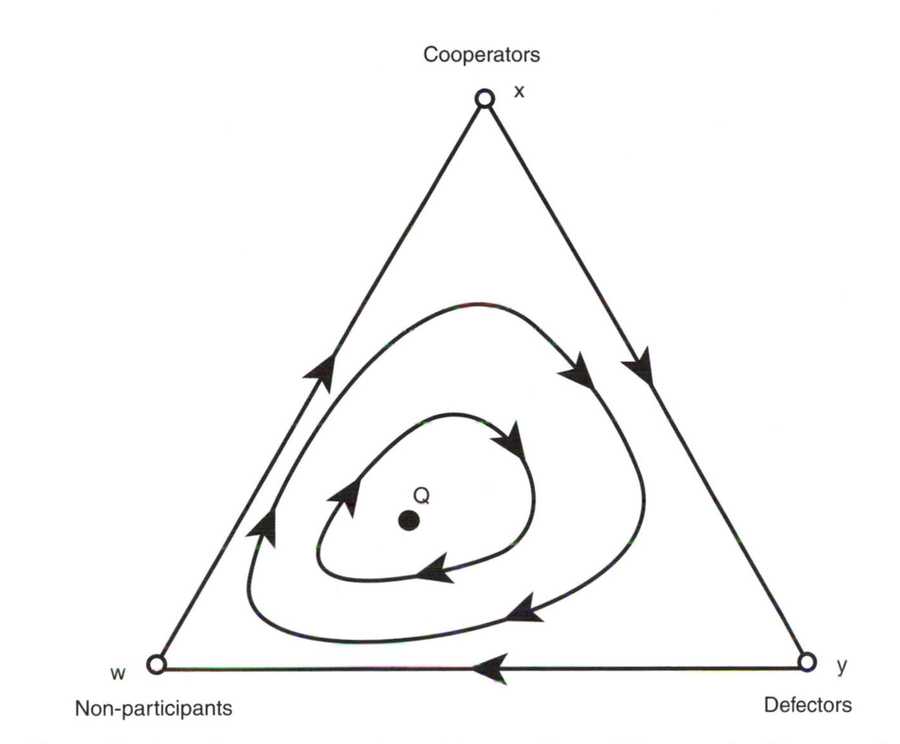

Figure 6.3 The replicator dynamics for the SR case, with $r > 2$. The rest point **Q** is surrounded by periodic orbits.

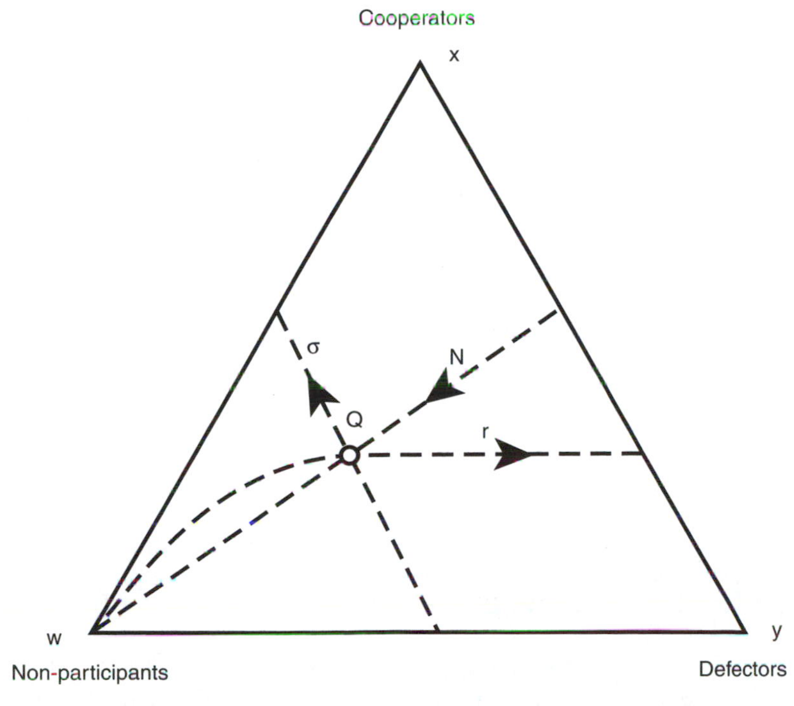

Figure 6.4 The dependence of **Q** on the parameters N, σ, and r.

which is independent of the sample size N. For increasing N, **Q** moves towards the vertex *AllW*. The limit $N \to \infty$ yields homoclinic orbits issuing from and leading to *AllW*.

6.9 TIME AVERAGES

Let us consider the replicator dynamics for the optional, self-returning Public Goods game. The time average of a function $v : S_3 \to R$ over an orbit of period T is defined as $\bar{v} = \frac{1}{T}\int_0^T v(x(t), y(t), w(t))dt$. It depends, in general, on the initial condition, but the time average of the fraction of cooperators among individuals participating in the game corresponds to its value at the equilibrium point **Q**:

$$\frac{\bar{x}}{\bar{x}+\bar{y}} = \frac{\sigma}{c(r-1)}. \tag{6.55}$$

Moreover, the time average of the fraction of cooperators among all participants in Public Goods games, i.e., $\bar{f}$, corresponds to the fraction of the averages:

$$\bar{f} = \frac{\sigma}{c(r-1)}. \tag{6.56}$$

Interestingly, an increase in r, the profitability of the public good, always favors defection, since it decreases the fraction f of cooperators among those actually participating in the Public Goods game.

In order to derive equation (6.55), we use the relation $x = f(1-w)$. Dividing both sides of the equation (6.49) by $w(1-w^{N-1})$, we get:

$$\int_0^T [\sigma(1-w) - c(r-1)x]\ dt = \int_0^T \frac{\dot{w}dt}{w(1-w^{N-1})} = p(w)\Big|_{w(0)}^{w(T)}, \tag{6.57}$$

with $p(w)$ being a primitive of $[w(1-w^{N-1})]^{-1}$. Since the orbits are closed, the last term vanishes and the proportionality between $\bar{x}$ and $1-\bar{w} = \bar{x}+\bar{y}$ follows. Equation (6.56) is obtained in the same way, if equation (6.49) is divided by $w(1-w)(1-w^{N-1})$.

Due to the properties of the replicator equation, the time averages of the payoffs for the three different strategies are equal and reduce to the payoff of non-participants, i.e., σ:

$$\bar{P}_x = \bar{P}_y = \bar{P}_w = \sigma. \tag{6.58}$$

Thus, in the long run, no one does better or worse than the non-participants.

6.10 ENTER THE PUNISHERS

Let us now assume, as in section 6.3, that participants in the game can also punish the cheaters in their group. Thus we consider four types of players: (1) the non-participants; (2) the cooperators, who participate and contribute, but do not punish; (3) the defectors, who participate, but neither contribute nor punish; and (4) the punishers, who participate, contribute, and punish the defectors in their group. We denote the relative frequencies of cooperators, defectors, punishers, and non-participants in the infinite population by x, y, z, and w. Their frequencies in a given random sample of size N are denoted by N_x, N_y, N_z, and N_w respectively (with $N_x + N_y + N_z + N_w = N$, and $S = N_x + N_y + N_z$ as the number of participants in the Public Goods game). In contrast to section 6.4, we do not allow for reputation effects.

Up to now, we have assumed that after each round of the Public Goods game, each punisher punishes each defector, and that each act of punishment costs γ to the punisher and β to the punished player, so that punishers have to pay γN_y and defectors βN_z. This is the so-called "peer-punishment" or "private punishment" case, and will be denoted by PP. Alternatively, we may consider the so-called "sanctioning institution" treatment. In this case, denoted by SI, each of the N_z punishers pays a fee γ *before* the Public Goods game is actually played; then, a fine of size βN_z is imposed on the group of defectors. (If there is no defector, no one will be fined and the fee for the punishment is lost; otherwise, the defectors share the fine equally, so that each defector pays $\beta N_z / N_y$.)

The total payoff is the sum of a Public Goods term and a punishment term, (which is 0 for non-participants and cooperators). In the PP case, the expected punishment terms are easily seen to be

$$-\beta z(N-1) \tag{6.59}$$

for the defectors and

$$-\gamma y(N-1) \tag{6.60}$$

for the punishers. For the SI case, the corresponding terms are

$$-\beta \frac{z}{y}[1-(1-y)^{N-1}] \tag{6.61}$$

for the defectors and

$$-\gamma(1-w^{N-1}) \tag{6.62}$$

for the punishers. The latter is clear: a punisher pays the fine γ, except if no one else in the sample wants to participate in the game. Now considering a defector, we see that if there are m other defectors in the sample, there remain $N-m-1$ players, who are punishers with a probability $z/(1-y)$ each. The probability that exactly k of them are punishers is

$$\binom{N-m-1}{k}\left(\frac{z}{1-y}\right)^k\left(1-\frac{z}{1-y}\right)^{N-m-1-k}, \tag{6.63}$$

in which case the fine is $k\beta/(m+1)$. If there are m other defectors, this yields a conditional expectation of

$$-\frac{\beta}{m+1}\left(\frac{z}{1-y}\right)(N-m-1). \tag{6.64}$$

Altogether, the punishment term is

$$-\sum_{m=0}^{N-1}\left(\frac{\beta z}{1-y}\right)\frac{N-m-1}{m+1}\binom{N-1}{m}y^m(1-y)^{N-m-1}. \tag{6.65}$$

Since the last term in the sum is 0, and since

$$\binom{N-1}{m}\frac{N-m-1}{m+1}=\binom{N-1}{m+1}, \tag{6.66}$$

the punishment term is

$$-\frac{\beta z}{y}\sum_{m=0}^{N-2}\binom{N-1}{m+1}y^{m+1}(1-y)^{N-1-(m+1)}, \tag{6.67}$$

which yields (6.61).

6.11 REPLICATOR DYNAMICS WITH PEER-PUNISHMENT

The payoff values are the sum of a Public Goods term and a punishment term. The punishment terms have been computed in the previous section.

In the SR case, the payoff stemming from the public good is given by σ for the non-participants, by

$$\sigma w^{N-1} + rc(x+z)H_N(w) \tag{6.68}$$

for the defectors, see equation (6.39), and

$$\sigma w^{N-1} + c(r-1)(1-w^{N-1}) - rcyH_N(w) \tag{6.69}$$

for the cooperators and for the punishers, see equation (6.41), with

$$H_N(w)=\frac{1}{1-w}\left(1-\frac{1-w^N}{N(1-w)}\right). \tag{6.70}$$

In the OO case, the payoff obtained from the public good is given by

$$\sigma w^{N-1} + \frac{rc(x+z)(1-w^{N-1})}{1-w} \tag{6.71}$$

for a defector. Cooperators and punishers obtain the same term from the public good, reduced by $c(1-w^{N-1})$.

Let us now consider the replicator dynamics for the OO-PP case. After removing the common term σw^{N-1} from all payoffs, we obtain for the expected payoff values of non-participants, defectors, cooperators, and punishers

$$P_w = (1 - w^{N-1})\sigma, \tag{6.72}$$

$$P_y = (1 - w^{N-1})\left(rc\frac{x+z}{1-w}\right) - \beta z(N-1), \tag{6.73}$$

$$P_x = (1 - w^{N-1})\left(rc\frac{x+z}{1-w} - c\right), \tag{6.74}$$

and

$$P_z = (1 - w^{N-1})\left(rc\frac{x+z}{1-w} - c\right) - \gamma y(N-1). \tag{6.75}$$

In the interior of the simplex S_4 there is no fixed point since $P_z < P_x$. Hence all orbits converge to the boundary. On the face $z = 0$, we find a Rock-Paper-Scissors game: non-participants are dominated by cooperators, who are dominated by defectors, who are dominated by non-participants. In the interior of this face, all orbits are homoclinic orbits, converging to the non-participant state if time converges to $\pm\infty$ (see fig. 6.2).

It is easy to see that punishers dominate non-participants, and that punishers and defectors form a bi-stable system if $c < \beta(N-1)$. The edge of cooperators and punishers ($y = w = 0$) consists of fixed points, those with

$$z > \frac{c}{\beta(N-1)} \tag{6.76}$$

are saturated and hence are Nash-equilibria.

Let us now turn to the SR-PP case. We compute the payoffs as $P_w = \sigma$,

$$P_y = \sigma w^{N-1} + rc(x+z)H_N(w) - \beta z(N-1), \tag{6.77}$$

$$P_x = \sigma w^{N-1} + (r-1)c(1 - w^{N-1}) - rcyH_N(w), \tag{6.78}$$

and

$$P_z = \sigma w^{N-1} + (r-1)c(1 - w^{N-1}) - rcyH_N(w) - \gamma y(N-1). \tag{6.79}$$

The $y = w = 0$ edge consists of fixed points, and all those with

$$z > \frac{c}{\beta(N-1)}\left(1 - \frac{r}{N}\right) \tag{6.80}$$

are Nash equilibria. On the face $z = 0$, the edges form a heteroclinic cycle. If $r \leq 2$, the dynamics looks just as in the OO-PP case. But for $r > 2$, the face $w = 0$ contains a fixed point Q that is surrounded by periodic orbits, as seen in fig. 6.3. This point **Q** is saturated, and hence is a Nash equilibrium. Indeed, at **Q** one has $P_x = (r-1)c(1 - w^{N-1}) + \sigma w^{N-1} - rcyH_N(w) = \bar{P} = P_w = \sigma$, and hence $P_z - \bar{P} = P_z - P_x = -\gamma(N-1)y < 0$. Moreover, any closed orbit o with period T in the face $z = 0$ attracts neighboring orbits from the interior of S_4, in the sense that the time average of the "transversal growth rate," i.e., of $P_z - \bar{P}$, is negative. This can be shown as before, by noting that the time-averages along o satisfy the equalities $\hat{P}_x = \hat{P}_y = \hat{P}_w = \hat{\bar{P}} = \sigma$, so that

$$\hat{P}_z - \hat{\bar{P}} = \hat{P}_z - \hat{P}_x = -\gamma(N-1)\frac{1}{T}\int_0^T y dt < 0. \tag{6.81}$$

The periodic orbit o is thus saturated in this sense, i.e., transversally stable, and even attracting. We note that for very large orbits the state spends most of the time close to *AllW*. The transversal eigenvalue at that point is 0.

Since $P_z < P_x$ in the interior of the state space, all orbits converge to the boundary. There are two sets on the boundary that attract interior orbits: on one hand the segment on the cooperator-punisher edge satisfying inequality (6.76), and on the other hand the face $z = 0$ filled with periodic orbits. Due to the degenerate dynamics (with continua of rest points and periodic orbits), the replicator dynamics is structurally unstable: the slightest perturbation can yield a different outcome. Hence, it has little predictive value. In such a situation, it is preferable to turn to stochastic processes in finite populations for an answer.

6.12 FINITE POPULATIONS

Let us consider a finite population of size M. By X, Y, Z, and W, we denote the number of cooperators, defectors, punishers, and non-participants (with $X+Y+Z+W = M$). The probability that a random sample of size N contains N_x cooperators, N_y defectors, N_w punishers, and N_z non-participants is given by the hypergeometric distribution

$$\frac{\binom{X}{N_x}\binom{Y}{N_y}\binom{Z}{N_z}\binom{W}{N_w}}{\binom{M}{N}}. \tag{6.82}$$

(Since we consider a finite population, we have to assume sampling without replacement.)

It is straightforward, if somewhat laborious, to compute the expected payoff values for each of these types. For instance, in the SR-PP case, these are given by $P_w = \sigma$,

$$P_y = \frac{W_{N-1}}{(M-1)_{N-1}}\sigma + H - \frac{Z}{M-1}(N-1)\beta, \tag{6.83}$$

$$P_x = \frac{W_{N-1}}{(M-1)_{N-1}}\sigma + H - F(W)c, \tag{6.84}$$

$$P_z = \frac{W_{N-1}}{(M-1)_{N-1}}\sigma + H - F(W)c - \frac{Y}{M-1}(N-1)\gamma, \tag{6.85}$$

where we used the expressions

$$W_N = W(W-1)\cdots(W-N+1) \tag{6.86}$$

etc., as well as

$$H = rc\frac{X+Z}{M-W-1}\left[1 - \frac{1}{N(M-W)}\left(M-(W-N-1)\frac{W_{N-1}}{(M-1)_{N-1}}\right)\right] \tag{6.87}$$

and

$$F(W) = 1 - \frac{r}{N}\frac{M-N}{M-W-1} + \frac{W_{N-1}}{(M-1)_{N-1}}\left(\frac{r}{N}\frac{W+1}{M-W-1} + r\frac{M-W-2}{M-W-1} - 1\right). \tag{6.88}$$

Using this, we can compute the transition probabilities for each learning process: since the population consists of 4 types, they yield a Markov chain with $\binom{M+3}{3}$ states (X, Y, Z, W). For reasonable values of M such as $M = 100$ or $M = 1000$, this is a rather large number. If there is no mutation, the population will end up in a homogeneous state, with all individuals being cooperators ($X = M$), or defectors ($Y = M$), etc. If we add a very small "innovation term" to the imitation process, as described in section 2.17, these states are no longer absorbing, and we can perform a separation of time scales. This "adiabatic" approach reduces the process to a Markov chain describing the transitions between the four homogeneous states *AllX*, *AllY*, *AllZ*, and *AllW*. The corresponding left eigenvector $(\pi_x, \pi_y, \pi_z, \pi_w)$ yields the stationary distribution of the four homogeneous states.

We have used this approach in section 6.4, by considering a very simple imitation process. In this section, we test the robustness of the result by using the Moran process as a learning rule (see sections 2.15 to 2.17).

According to equation (2.78), the fixation probabilities can be derived by formulas of the type

$$\rho_{xy} = \frac{1}{1+\sum_{k=1}^{M-1}\prod_{X=1}^{k}\frac{1-s+sP_{XY}}{1-s+sP_{YX}}}, \tag{6.89}$$

where P_{XY} denotes the payoff obtained by a cooperator in a population consisting of X cooperators and $Y = M - X$ defectors. Hence all that remains is to compute these expressions.

6.13 PAYOFF VALUES IN A FINITE POPULATION

The payoff consist of two terms: the term G from the Public Goods game, and the term S from the sanctioning. We consider only the case of a self-returning Public Goods game with peer-punishment.

In the PP case, punishment affects only punishers and defectors. If there are Z punishers and Y defectors (with $Z + Y = M$), the former must pay

$$S_{ZY} = \frac{\gamma Y(N-1)}{M-1}, \tag{6.90}$$

and the latter

$$S_{YZ} = \frac{\beta Z(N-1)}{M-1}. \tag{6.91}$$

All other S terms are 0 in the PP case.

If the population consists only of X cooperators and Y defectors ($X + Y = M$), the payoff obtained from the Public Goods interaction is

$$G_{XY} = \sum_{k=0}^{N-1} H(k, N-1, X-1, M-1)\left(\frac{k+1}{N}r - 1\right)c, \quad (6.92)$$

where

$$H(k, N-1, X-1, M-1) = \frac{\binom{X-1}{k}\binom{M-X}{N-k-1}}{\binom{M-1}{N-1}} \quad (6.93)$$

denotes the probability that a cooperator lands in a sample with k other cooperators and $N - 1 - k$ defectors, cf. expression (6.81). Thus

$$G_{XY} = \frac{rc}{N}\left[1 + (X-1)\frac{N-1}{M-1}\right] - c. \quad (6.94)$$

Similarly,

$$G_{YX} = \left(\frac{rc}{N}\right)\frac{N-1}{M-1}X, \quad (6.95)$$

$$G_{XZ} = G_{ZX} = (r-1)c, \quad (6.96)$$

$$G_{WX} = G_{WY} = G_{WZ} = \sigma, \quad (6.97)$$

$$G_{XW} = G_{ZW} = (r-1)c - \frac{W_{N-1}}{(M-1)_{N-1}}((r-1)c - \sigma), \quad (6.98)$$

$$G_{YW} = \frac{W_{N-1}}{(M-1)_{N-1}}\sigma, \quad (6.99)$$

$$G_{YZ} = \frac{Z(N-1)}{M-1}\left(\frac{rc}{N}\right), \quad (6.100)$$

and

$$G_{ZY} = \frac{rc}{N}\left[\frac{(Z-1)(N-1)}{M-1} + 1\right] - c. \quad (6.101)$$

It is now possible to compute the stationary distribution for the Markov chain, both in the case when participation is voluntary, and when it is compulsory. We see that in the former case, punishers dominate, see figure 6.5, whereas in the latter case, defectors win, see figure 6.6. Individual based simulations illustrate this point. Cooperation based on sanctioning is much easier to achieve if participation in the joint effort is voluntary, rather than compulsory, see figure 6.7. Similar results also hold for the other cases (i.e., a Public Goods game of "others-only" type, or a punishment term provided by a "sanctioning institution").

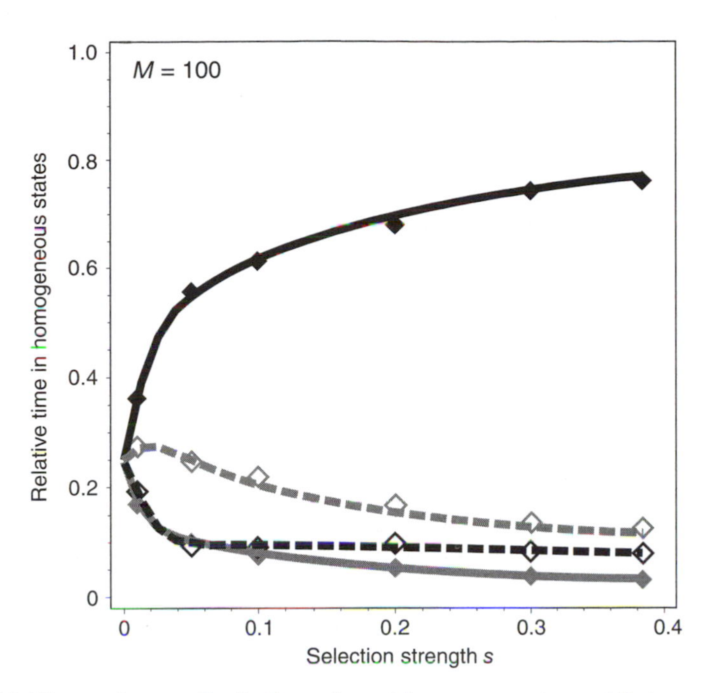

Figure 6.5 The stationary distribution of punishers, cooperators, defectors, and non-participants (in that order, from top to bottom) for various values of the selection strength s. Parameter values $M = 100$, $N = 5$, $r = 3$, $\sigma = 1$, $\beta = 1$, $\gamma = 0.3$, and $c = 1$.

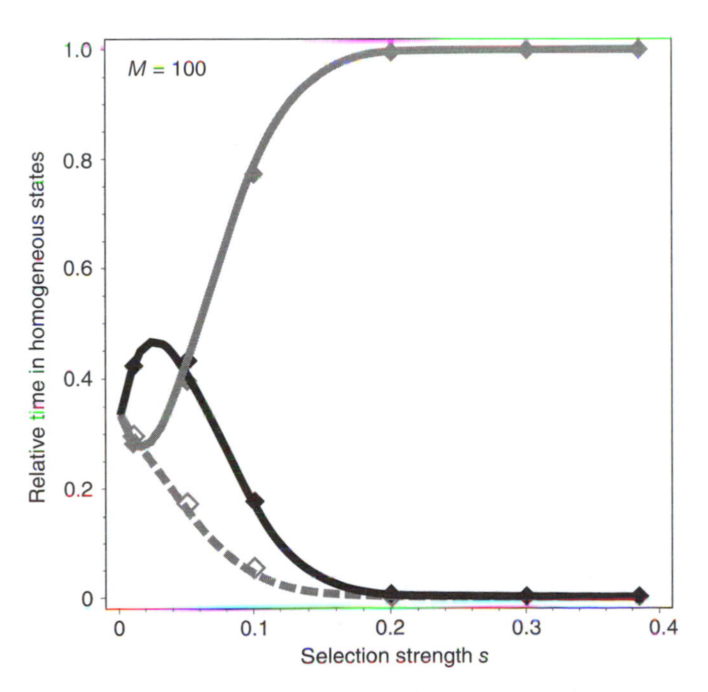

Figure 6.6 The stationary distribution for the compulsory game (all players must participate), for the same parameter values. The ordering is now (from top to bottom, right edge): defectors, punishers, and cooperators.

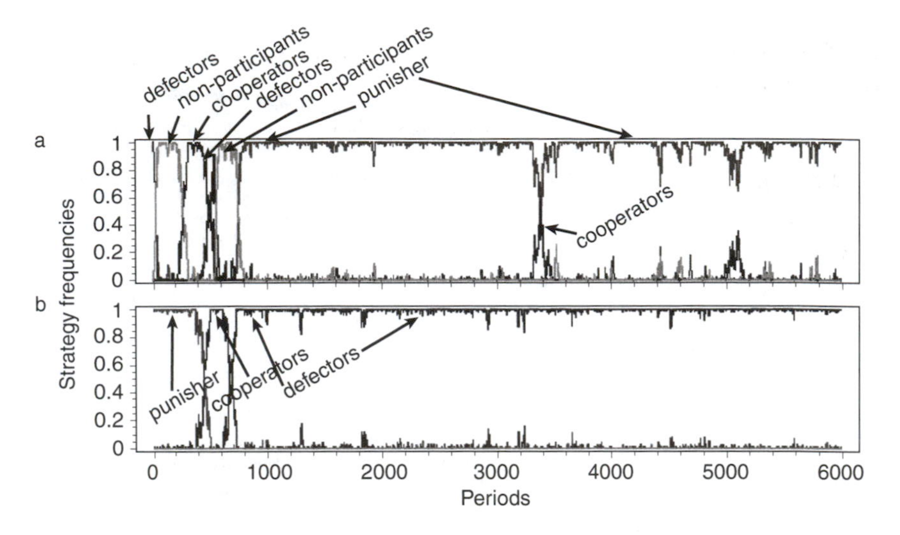

Figure 6.7 A sketch of a typical evolution of (a) the voluntary Public Goods game with punishment, and (b) the compulsory Public Goods game with punishment. For corresponding individual-based computer simulations, see Hauert et al. (2007).

6.14 REFERENCES

That players can refuse to take part in a game was stressed by Orbell and Dawes (1993), see also Battali and Kitcher (1995). Our treatment of voluntary Public Goods games is taken from Hauert et al. (2002a, 2002b), see also Hauert, Haiden, and Sigmund (2004), Castro and Toro (2008) and, for an experimental test, Semmann, Krambeck, and Milinski (2003). The importance of non-participation for the emergence of punishment was first proposed by Fowler (2005a), see also Brandt, Hauert, and Sigmund (2006) and Boyd and Mathew (2007). The approach via finite populations is from Hauert et al. (2007), see also Traulsen et al. (2009) and De Silva et al. (2009). For online simulations, see the homepages of Hannelore De Silva, http://www.wu-wien.ac.at/usr/ma/hbrandt/publicgoods/ and of Christoph Hauert, http://www.univie.ac.at/virtuallabs/. There exists a large literature on punishment, we refer to Boyd and Richerson (1992), Sell and Wilson (1999), Henrich and Boyd (2001), Boyd et al. (2003), Masclet et al. (2003), Falk, Fehr, and Fischbacher (2005), Small and Loewenstein (2005), Xiao and Houser (2005), and Dawes et al. (2007). Some authors considered second-order punishment (inflicted on those who do not punish defectors), but it affects the emergence of sanctioning only marginally, and seems to have little empirical support, see Kiyonari et al. (2004) and Fowler (2005b). Johnson and Bering (2006) argue that the fear of supernatural punishment plays an important role in the emergence of cooperation. For the impact of social framing, see Rege and Telle (2004). Various counter-productive aspects of punishment are displayed in experiments by Gneezy and Rustichini (2000) or Fehr and Rockenbach (2003). For interactions between Public Goods games and indirect reciprocity, see Milinski, Semmann, and Krambeck (2002b), Semmann, Krambeck, and Milinski (2004), Rockenbach and Milinski (2006), and Panchanathan and Boyd (2006).

Chapter Seven

Cooperation in Structured Populations

7.1 STRUCTURED POPULATIONS

Up to now, we have always assumed that populations are well-mixed: just as population genetics often postulates *random mating*, so did we assume *random meeting*. It need hardly be stressed that this is often unrealistic. Human populations are, in general, highly structured, and individuals interact preferentially within their families, neighborhoods, or other types of groups and networks. Obviously, these structures play a major role in the evolution of cooperation.

So far, we have concentrated exclusively on an analysis based on the strategic point of view, and totally neglected social networks, despite their obvious importance. In this last chapter, we briefly discuss some relevant aspects, not to make belated amends for an oversight, but to point out some of the main directions.

7.2 KIN SELECTION

A huge part of cooperation occurs within families. This is an immediate corollary of the Darwinian struggle for survival. Genes that promote their own spreading (by enhancing the survival and the fecundity of their carriers) become necessarily more frequent than those that do not. Just as parents programmed to help their children have an obvious advantage in passing along their genetic material, so siblings programmed to help each other will also have an advantage. More precisely, a gene causing me to help my brother will help to spread itself: for it is, with a high probability, carried by my brother too. This "selfish gene" view, elaborated as the theory of kin selection, has been developed to a considerable extent, within the last fifty years. Here, we just derive an elementary instance of that theory's basic result, which is known as *Hamilton's rule*.

In the following, we shall interpret "payoff" in the sense of genetic fitness, i.e., reproductive success. Let ρ denote the *coefficient of relatedness* between two players. This can be defined in various ways. Here, we simply assume that it measures *relatedness by descent*: this is the probability that a recently mutated gene (or allele, to use the correct term) carried by one player is also carried by the other. Of course, any two humans are related, if we go back to primordial Eve. But we do not share all our genes. A mutation occurring in the body of your grandfather produces an allele that will be found with probability $1/2$ in his children, and with probability $1/4$ in his grandchildren. Under usual circumstances (e.g., no inbreeding), the coefficient

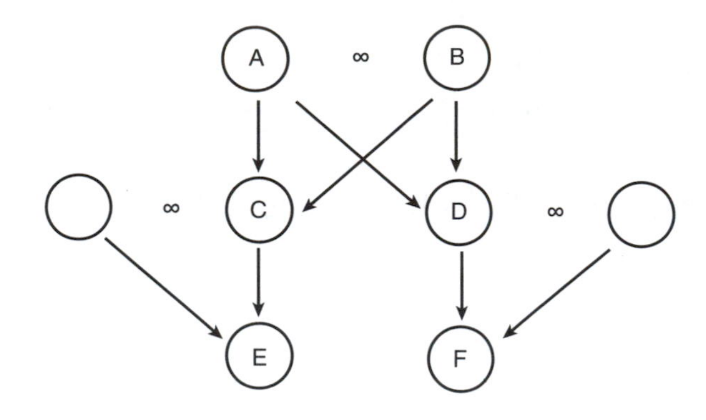

Figure 7.1 Relatedness in a small family. A and B are the parents, C and D their offspring, E and F their grandchildren. Individuals have two copies (alleles) of each gene, one inherited from the mother, the other from the father. The probability that a specific allele of A is passed to C is 1/2. The probability that a specific allele of C comes from A is 1/2. The degree of relatedness between A and C is 1/2. The siblings C and D can both have inherited a newly mutated allele, either from A (probability $1/2 \times 1/2 = 1/4$) or from B, hence their degree of relatedness is $1/4 + 1/4 = 1/2$. The degree of relatedness between F and C (nephew and uncle) is $1/2 \times 1/2 \times 1/2 + 1/2 \times 1/2 \times 1/2 = 1/4$, the degree of relatedness between E and F (two cousins) is 1/8.

of relatedness between two siblings is $1/2$; between you and your nephew it is $1/4$, etc., see figure 7.1.

From the gene's point of view, this means that the reproductive success of a relative also counts towards your own reproductive success, but weighted by the factor $\rho < 1$, as it contributes to your indirect fitness. Assume that you are engaged in a Donation game, with matrix

$$\begin{pmatrix} b-c & -c \\ b & 0 \end{pmatrix}. \tag{7.1}$$

Clearly, the second strategy dominates the first (see section 3.1). But if the payoff of your co-player, multiplied by the factor ρ, is added to your own, then the payoff matrix turns into

$$\begin{pmatrix} (1+\rho)(b-c) & -c+b\rho \\ b-c\rho & 0 \end{pmatrix}. \tag{7.2}$$

It is easy to see that $\rho b > c$ implies that it is now the first strategy (i.e., cooperation) that dominates: no matter whether your co-player defects or cooperates, it will be better for you to cooperate. For full siblings, for instance, this condition means that the benefit has to be larger than twice the cost. The relation

$$\rho > \frac{c}{b} \tag{7.3}$$

is known as Hamilton's rule. If it is satisfied, then it pays to cooperate. Hamilton derived this rule in the context of population genetics. But clearly, it holds whenever it can be assumed that the utility of your co-player (properly weighted by some factor ρ) counts toward your own utility. This need not be restricted to games between relatives. It may well hold simply because you like your co-player. (Of course, a proper Darwinian would then require an explanation for the genetic basis of your sympathy.)

We have encountered inequalities similar to Hamilton's rule before. In particular, in the theory of direct reciprocity (section 3.4) we have met the important inequality

$$w > c/b \tag{7.4}$$

(where w is the probability for a further round), and in the theory of indirect reciprocity (section 4.3) the inequality

$$q > c/b \tag{7.5}$$

(where q is the probability to know the co-player's previous move).

7.3 GAMES ON GRIDS

Let us now consider another scenario. Assume that players are not randomly milling around, but sedentary. They interact only with their neighbors. In the simplest case, let us assume that they are living on a huge chessboard, each on his or her own site. Then each player has eight neighbors, and interacts only with those. We shall assume that the players have to engage in the Donation game, with the payoff matrix (7.1), and that they always use the same strategy, i.e., always play C or always D, with each of their eight neighbors. The payoff for a defector is $N_c b$, where $0 \leq N_c \leq 8$ is the number of cooperating neighbors. The payoff for a cooperator is $N_c b - 8c$. Suppose that players play one round of the Donation game with each of the eight neighbors, and then update strategy, by adopting the strategy of the neighbor with the highest payoff, or by sticking to their own strategy, if it worked best. In the case of a tie, we assume some random decision.

It is easy to see that a single cooperator in a sea of defectors will vanish, whereas a single defector will infect all eight neighbors. But for larger clusters, the outcome may depend on the precise geometry, see figure 7.2. For instance, when a straight line separates cooperators from defectors, then front-line cooperators earn $5b - 8c$ and front-line defectors earn $3b$. Thus cooperators do better than defectors if $b/c > 4$. Their front will advance if this condition is met. A cooperator in the second row earns $8b - 8c$, and the defectors' front will advance if $b/c < 8/5$. On the other hand, a cooperator sitting at the corner of a block of cooperators earns $3b - 8c$, while some defecting neighbors obtain $3b$, which is always more. But a defector sitting at the corner of a block of defectors earns $5b$, and this is less than the payoff $7b - 8c$ obtained by the cooperator who sits in the diagonally opposite site, provided $b > 4c$.

Of course, the distribution of defectors and cooperators is unlikely to be given by a simple geometric configuration. More generally, we cannot realistically assume that the players sit on a regular lattice, such as an infinite chessboard. The neighborhood relations will usually be more random, and given by an irregular graph.

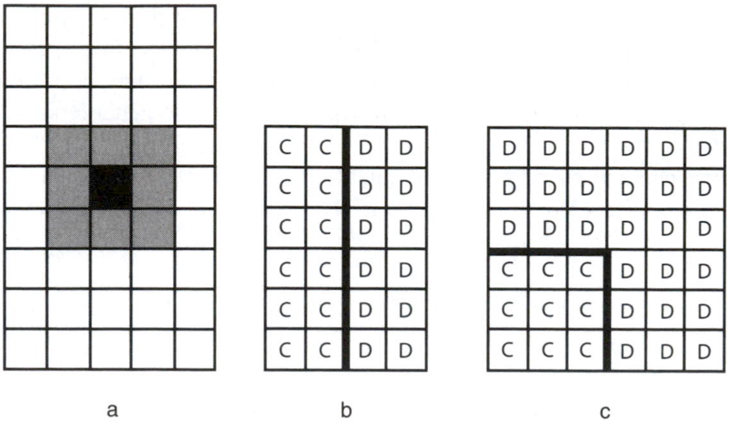

Figure 7.2 Games on grids: (a) describes the neighborhood of an individual on a specific site; (b) depicts a straight front-line between cooperators and defectors; (c) describes the corner of a rectangular block of cooperators in a sea of defectors.

Furthermore, there are many alternative ways of modeling how strategies spread, through imitation or otherwise. Players can update asynchronously, they can use some stochastic learning rules, etc. But as long as the interaction network is not too irregular, and players imitate their neighbors with a probability proportional to their fitness, and this fitness depends only weakly on the payoff, then a rule of thumb states that cooperation wins whenever

$$k < \frac{b}{c}, \tag{7.6}$$

where k is the average number of neighbors. Intuitively, this makes sense. The more neighbors, the less special they are, and the closer the population is to being well-mixed.

7.4 THE PRICE EQUATION

Whenever a population is subdivided into subpopulations, the Price equation offers a useful tool for computing evolutionary change. Let us assume, as sketched in figure 7.3, that the subpopulation i (for $1 \leq i \leq m$) has size N_i, and that it displays a certain trait p_i (for instance, the frequency of cooperators in the subpopulation, or the total body mass of its members). The total population size then, is $N = \sum N_i$ and the average trait value is $E(p) = \sum N_i p_i / N$.

Let us assume that the individuals multiply, so that in the next generation, the subpopulations' sizes have become N_i' and their trait values p_i'. The average trait value is now $E'(p') = \sum N_i' p_i' / N'$. (The symbol E' emphasizes that the average is taken with respect to the weights of the subpopulations in the *new* generation.) The fractions N_i'/N_i can be viewed as the per capita reproduction, or average fitness in group i, and will be denoted by f_i. The Price equation establishes a relation between

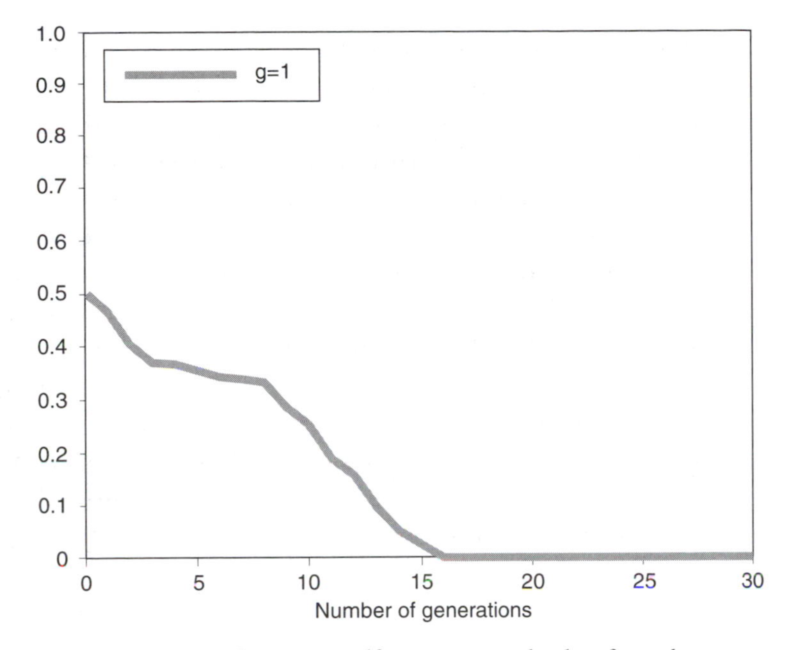

Figure 7.3 The frequency of cooperators if groups are randomly reformed every generation.

$\Delta p = E'(p') - E(p)$, the increase of the average trait value in the whole population, and the increases $\delta p_i = p'_i - p_i$ of these traits in each subpopulation. In general, the difference of the averages is distinct from the average of the differences. But the two expressions can be related, using the covariance of the two functions $p : i \mapsto p_i$ and $f : i \mapsto f_i$, i.e.,

$$cov(f, p) = E(fp) - E(f)E(p). \tag{7.7}$$

Here, E denotes the mean, as before, i.e.,

$$E(f) = \frac{1}{N} \sum N_i f_i, \tag{7.8}$$

and

$$E(fp) = \frac{1}{N} \sum N_i f_i p_i. \tag{7.9}$$

Clearly,

$$E(f) = \frac{N'}{N} \tag{7.10}$$

and

$$E(f\delta p) = \frac{1}{N} \sum N_i f_i \delta p_i = E(fp') - E(fp). \tag{7.11}$$

Since

$$E(fp') = \frac{1}{N}\sum f_i p_i' N_i = \frac{1}{N}\sum \frac{N_i'}{N_i} p_i' N_i = \frac{N'}{N}\sum \frac{N_i' p_i'}{N'} = E(f)E'(p'), \tag{7.12}$$

the expression (7.7) for the covariance yields the *Price equation*

$$E(f)[E'(p') - E(p)] = cov(f, p) + E(f\delta p) \tag{7.13}$$

or, if $N' > 0$, i.e., if $E(f) \neq 0$,

$$\Delta p = \frac{cov(f, p)}{E(f)} + \frac{E(f\delta p)}{E(f)}. \tag{7.14}$$

The term $cov(f, p)/E(f)$ may be interpreted as

$$\begin{aligned} \frac{E(fp)}{E(f)} - \frac{E(f)E(p)}{E(f)} &= \frac{\sum N_i f_i p_i/N}{N'/N} - E(p) \\ &= \sum N_i' p_i/N' - E(p) = E'(p) - E(p). \end{aligned} \tag{7.15}$$

Here, $E'(p)$ can be understood as the average value of the p_i (the trait values in the initial generation), if the groups had the sizes attained *in the following generation.*

7.5 ADVANCE OF ALTRUISTS

Let us apply this to the Public Goods game, and assume that group i contains X_i cooperators and Y_i defectors, with $X_i + Y_i = N_i$. The frequency of cooperators is $p_i = X_i/N_i$. Let us suppose that within each group, a Public Goods game is played. If we assume that it is of the self-returning type (see section 6.2), then we obtain as payoffs P_d and P_c for the defectors and the cooperators:

$$P_d(i) = rcX_i/N_i = rcp_i \tag{7.16}$$

(with $r > 1$), and

$$P_c(i) = rcp_i - c. \tag{7.17}$$

Let the fitness function be a convex combination of a baseline term $B > 0$ (the same for all), and the payoff from the game. Then the fitness values for the defectors and cooperators in group i are given by $(1-s)B + sP_d(i)$ and $(1-s)B + sP_c(i)$, for some "selection coefficient" $s \in [0, 1]$. (The main use of s is to make sure that the fitness terms are positive. This can always be guaranteed by choosing s small enough.) The average fitness in group i is

$$f_i = (1-s)B + s[rcp_i(p_i + (1-p_i)) - cp_i] = (1-s)B + sc(r-1)p_i. \tag{7.18}$$

The right hand side of the Price equation (7.13) is the sum of two terms. The second term can never be positive. Indeed, within each group, the defectors do better than the cooperators, and hence the expected frequencies of the cooperators will drop: $\delta p_i \leq 0$ for all i. But the first term is non-negative. Indeed,

$$cov(f, p) = sc(r-1)cov(p, p) = sc(r-1)var(p) \geq 0. \tag{7.19}$$

Hence, if the variance of p is sufficiently large, i.e., if the subgroups differ sufficiently in their composition, then the average frequency of cooperators in the whole population can increase from one generation to the next. (This also holds, incidentally, for the others-only case of the Public Goods game.) While defectors spread within each group, those groups having few defectors grow faster and thus can compensate.

It must be stressed that this is a fleeting effect only. Over the generations, groups will become more and more homogeneous (since each one will be more and more dominated by defectors), hence the variance will decrease and ultimately no longer suffice to deliver a net increase in cooperators, see figure 7.3.

However, this can be overcome in various ways: for instance, by an appropriate migration between the groups. Another way would be to randomly regroup the population every few generations: if the number of generations is chosen judiciously within a certain range, the variance can rebound again and again, see figure 7.4. We can also assume that cooperators will preferentially unite with other cooperators. Defectors would also prefer to join with cooperators, of course, but as long as the others have a say, defectors can be kept away.

Another, more drastic but not entirely unrealistic approach would be to eliminate those groups with many defectors, and to replace them with groups obtained by randomly splitting groups with many cooperators. We can assume, for instance, that groups with many defectors are vanquished in warfare by groups containing more altruists ready to risk their own lives in battle. This approach can trace its roots back to Darwin, who wrote: "There can be no doubt that a tribe including many members who . . . were always ready to give aid to each other and to sacrifice themselves for the common good, would be victorious over other tribes; and this would be natural selection."

7.6 ANOTHER VERSION OF HAMILTON'S RULE

A related interpretation of the role of the variance in p can be obtained by looking for a necessary and sufficient condition for an increase in the frequency of cooperators. Let us denote by x_i' and y_i' the expected values of X_i' and Y_i', the frequencies of cooperators and defectors in the next generation. Since

$$x_i' = X_i[(1-s)B + sP_c(i)] \tag{7.20}$$

and

$$y_i' = Y_i[(1-s)B + sP_d(i)], \tag{7.21}$$

(and using $x' = \sum x_i'$ and $y' = \sum y_i'$ as well as $n' = x' + y'$), the condition

$$x'/n' > X/N \tag{7.22}$$

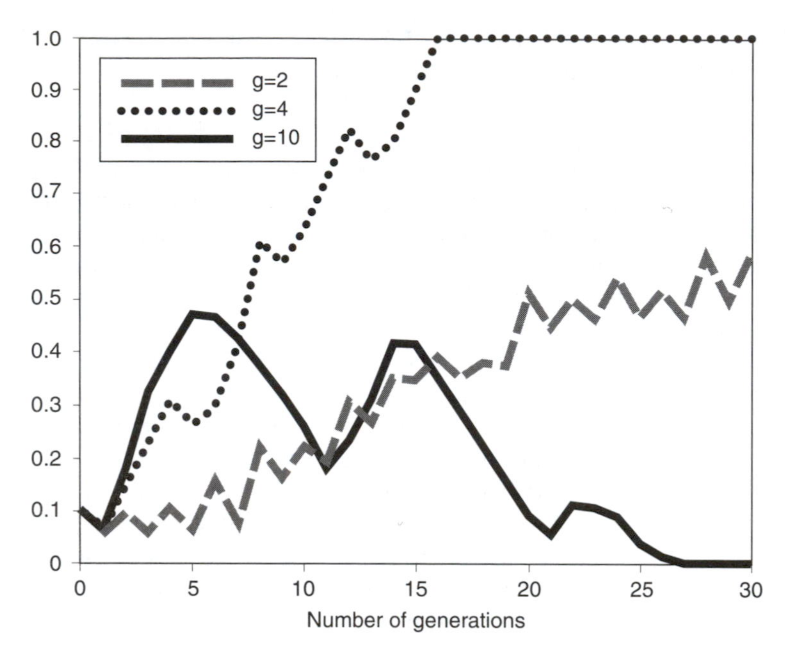

Figure 7.4 The frequency of cooperators if groups are randomly reformed every second, fourth, or tenth generation (after Pflügl (2007), based on Fletcher and Zwick (2004)). The length of the period g between the new arrangements plays a crucial role for the evolution of cooperation.

translates into

$$\frac{(1-s)BX + s\sum X_i P_c(i)}{(1-s)BN + s(\sum X_i P_c(i) + \sum Y_i P_d(i))} > \frac{X}{N}. \tag{7.23}$$

This in turn is equivalent to

$$N\sum X_i P_c(i) > X\left(\sum X_i P_c(i) + \sum Y_i P_d(i)\right). \tag{7.24}$$

Using equations (7.16) and (7.17), and collecting the terms in r on the left hand side, we obtain

$$r\left[N\sum X_i p_i - X\sum N_i p_i\right] > X(N-X). \tag{7.25}$$

After division by N^2, this gives

$$r\left[\sum p_i \frac{X_i}{N} - E(p)\sum p_i \frac{N_i}{N}\right] > E(p)(1-E(p)). \tag{7.26}$$

The factor of r can be written as

$$\frac{1}{N}\sum p_i^2 N_i - E(p)\frac{1}{N}\sum N_i p_i = E(p^2) - (E(p))^2 = var(p). \tag{7.27}$$

Hence the proportion of cooperators increases if and only if,

$$r[var(p)] > E(p)(1 - E(p)). \tag{7.28}$$

Let us denote the right hand side by $Var(p)$. It is the variance of the random variable obtained by sampling one individual at random in the whole population, and checking whether it cooperates (value 1) or defects (value 0). If we recall that c is the cost of an individual contribution to the public good, and $rc =: b$ the benefit accruing to the whole subpopulation, we obtain

$$\frac{var(p)}{Var(p)} > \frac{c}{b}, \tag{7.29}$$

which is another inequality of Hamiltonian type. It highlights the importance of the variance of p. If the proportions of cooperators in the different subpopulations are too similar, the proportion of cooperators in the whole population decreases.

7.7 REFERENCES

A unifying approach to cooperation in structured und unstructured populations has been proposed by Nowak (2006b). The founding father of the theory of kin selection was W. D. Hamilton, see Hamilton (1996) for a reprint of his seminal papers and Grafen (1984) for a lucid presentation. For modern presentations, we refer to Frank (1998), Taylor and Frank (1996), Pepper (2000), Taylor, Day, and Wild (2007), Skyrms and Pemantle (2000), West, Pen, and Griffin (2002), Rousset (2004), and Lehmann and Keller (2006). Games on grids were introduced by Nowak and May (1992), see also Nowak, Bonhoeffer, and May (1994), Lindgren and Nordahl (1994), and Hauert and Szabó (2003). The importance of spatial relations has been stressed by Schelling (1971) and Durrett and Levin (1994). Reaction-diffusion equations for the Prisoner's Dilemma game were studied in Hutson and Vickers (1995) and Ferrière and Michod (1996). For cooperative games on irregular networks and graphs, we refer to Liebermann, Hauert, and Nowak (2005), Szabó and Fáth (2007), Santos, Pacheco, and Lenaerts (2006), and Pacheco et al. (2008). For the spatial Ultimatum game see Page, Nowak, and Sigmund (2000); for spatial indirect reciprocity see Brandt et al. (2007); for the spatial Snowdrift game, Hauert and Doebeli (2004) and Doebeli, Hauert, and Killingback (2004); for the spatial Public Goods game, Brandt, Hauert, and Sigmund (2003), Nakamaru and Iwasa (2005 and 2006); for the effect of volunteering, Szabó and Hauert (2002) and Szabó and Vukov (2004). The Price equation was introduced in Price (1970), see also Grafen (2000) and, for an intriguing relation with the replicator equation, Page and Nowak (2002). The Price equation has been the main basis of a large body of work; see for instance Frank (1998), Sober and Wilson (1998), or Gintis (2000). The presentation in sections 7.5 and 7.6, as well as figure 7.3 are based on Fletcher and Zwick (2004 and 2007). Different approaches to group selection can be found in Cohen and Eshel (1976), Bergstrom (2002), or Killingback Bieri, and Flatt (2006). Lehmann and Keller (2006) and Lehmann et al. (2007) argue that group selection is kin selection, see also West, Griffin, and Gardner (2007).

References

Abreu, D., and A. Rubinstein. 1988. "The structure of Nash equilibria in repeated games with finite automata." *Econometrica* 56:1259–82.

Alexander, R. D. 1987. *The Biology of Moral Systems*. New York: Aldine de Gruyter.

Andreoni, J., W. T. Harbaugh, and L. Vesterlund. 2003. "The carrot or the stick: Rewards, punishments, and cooperation." *American Economic Review* 93:893–902.

Ashrav, N., I. Bohnet, and N. Pinkov. 2006. "Decomposing trust and trustworthiness." *Experimental Economics* 9:193–208.

Aumann, R. 1981. "Survey of repeated games." In *Essays in Game Theory and Mathematical Economy in Honor of Oskar Morgenstern*. Mannheim: Wissenschaftsverlag, Bibliographisches Institut.

Axelrod, R. 1984. *The Evolution of Cooperation*. New York: Basic Books.

Axelrod, R., and D. Dion. 1988. "The further evolution of cooperation." *Science* 242:1385–90.

Axelrod, R., and W. D. Hamilton. 1981. "The evolution of cooperation." *Science* 211:1390–96.

Banks, J. S., and R. K. Sundaram. 1990. "Repeated games, finite automata and complexity." *Games and Economic Behaviour* 2:97–117.

Bateson, M., D. Nettle, and G. Roberts. 2006. "Cues of being watched enhance cooperation in a real-world setting." *Biology Letters* 2:412–14.

Battali, J., and P. Kitcher. 1995. "Evolution of altruism in optional and compulsory games." *Journal of Theoretical Biology* 175:161–71.

Baumeister, R. F., E. Bratlavsky, C. Finkenauer, and K. D. Vohs. 2001. "Bad is stronger than good." *Review of General Psychology* 5:323–70.

Bendor, P., and J. Swistak. 1995. "Types of evolutionary stability and the problem of cooperation." *Proceedings of the National Academy of Sciences* 92:3596–3600.

Bendor, P., and J. Swistak. 2001. "The evolution of norms." *American Journal of Sociobiology* 106:1493–1545.

Berg, J., J. Dickhaut, and K. McCabe. 1995. "Trust, reciprocity, and social history." *Games and Economic Behavior* 10:122–42.

Bergstrom, T. C. 2002. "Evolution of social behaviour: Individual and group selection." *Journal of Economic Perspectives* 16:67–88.

Binmore, K. G. 1994. *Playing Fair: Game Theory and the Social Contract*. Cambridge, MA: MIT Press.

Binmore, K. G., and L. Samuelson. 1992. "Evolutionary stability in repeated games played by finite automata." *Journal of Economic Theory* 57:278–305.

Boerlijst, M. C., M. A. Nowak, and K. Sigmund. 1997a. "The logic of contrition." *Journal of Theoretical Biology* 185:281–94.

Boerlijst, M. C., M. A. Nowak, and K. Sigmund. 1997b. "Equal pay for all prisoners." *American Mathematical Monthly* 104:303–7.

Bohnet, I., and R. Croson. 2004. "Introduction to special issue on trust and trustworthiness." *Journal of Economic Behavior and Organization* 55:443–45.

Bolton, G., E. Katok, and A. Ockenfels. 2004a. "How effective are online reputation mechanisms? An experimental investigation." *Management Science* 50:1587–1602.

Bolton, G., E. Katok, and A. Ockenfels. 2004b. "Cooperation among strangers with limited information about reputation." *Journal of Public Economics* 89:1457–68.

Bolton, G., and A. Ockenfels. 2000. "ERC—A theory of equity, reciprocity and competition." *American Economic Review* 90:166–93.

Bomze, I. 1983. "Lotka-Volterra equations and replicator dynamics: A two-dimensional classification." *Biological Cybernetics* 48:201–11.

Bowles, S., and H. Gintis. 2002. "Homo reciprocans." *Nature* 415:125–28.

Boyd, R. 1989. "Mistakes allow evolutionarily stability in the repeated Prisoner's Dilemma game." *Journal of Theoretical Biology* 136:47–59.

Boyd, R., H. Gintis, S. Bowles, and P. J. Richerson. 2003. "The evolution of altruistic punishment." *Proceedings of the National Academy of Sciences* 100:3531–35.

Boyd, R., and S. Mathew. 2007. "A narrow road to cooperation." *Science* 316:1858–59.

Boyd, R., and P. J. Richerson. 1989. "The evolution of indirect reciprocity." *Social Networks* 11:213–36.

Boyd, R., and P. J. Richerson. 1992. "Punishment allows the evolution of cooperation (or anything else) in sizeable groups." *Ethology and Sociobiology* 113:171–95.

Brandt, H., C. Hauert, and K. Sigmund. 2003. "Punishment and reputation in spatial public goods games." *Proceedings of the Royal Society B* 270:1099–1104.

Brandt, H., C. Hauert, and K. Sigmund. 2006. "Punishing and abstaining for public goods." *Proceedings of the National Academy of Sciences* 103:495–97.

Brandt, H., and K. Sigmund. 2004. "The logic of reprobation: Assessment and action rules for indirect reciprocity." *Journal of Theoretical Biology* 231:475–86.

Brandt, H., and K. Sigmund. 2005. "Indirect reciprocity, image scoring, and moral hazard." *Proceedings of the National Academy of Sciences* 102:2666–70.

Brandt, H., and K. Sigmund. 2006. "The good, the bad and the discriminator: Errors in direct and indirect reciprocity." *Journal of Theoretical Biology* 239:183–94.

Brandt, H., K. Sigmund, H. Ohtsuki, and Y. Iwasa. 2007. "A survey on indirect reciprocity." In Takeuchi, Y., Y. Iwasa, and K. Sato, eds. *Mathematics for Ecology and Environmental Sciences*. Berlin: Springer.

Brosnan, S. F., and F.B.M. de Waal. 2003. "Monkeys reject unequal pay." *Nature* 425:297–99.

Brown, D. 1991. *Human Universals*. McGraw-Hill.

Bshary, R., and A. S. Grutter. 2005. "Punishment and partner switching causes cooperative behavior in a cleaning mutualism." *Biology Letters* 1:396–99.

Bshary, R., and A. S. Grutter. 2006. "Image scoring causes cooperation in a cleaning mutualism." *Nature* 441:975–78.

Burnham, T., and B. Hare. 2007. "Engineering cooperation: Does involuntary neural activation increase public goods contributions?". *Human Nature* 18:88–107.

Burnham, T., and D.D.P. Johnson. 2005. "The biological and evolutionary logic of human cooperation." *Analyse und Kritik* 27:113–35.

Camerer, C. 2003. *Behavioral Game Theory: Experiments in Strategic Interactions*. Princeton University Press.

Camerer, C., and E. Fehr. 2006. "When does 'economic man' dominate social behaviour?" *Science* 311:47–52.

Carpenter, J., G. W. Harrison, and J. A. List, eds. 2005. *Field Experiments in Economics*. Vol. 10. Research in Experimental Economics. Amsterdam: Elsevier.

Castro, L., and M. Toro. 2008. "Iterated Prisoner's Dilemma in an asocial world dominated by loners, not defectors." *Theoretical Population Biology* 74:1–5.

Chalub, F., F. C. Santos, and J. M. Pacheco. 2006. "The evolution of norms." *Journal of Theoretical Biology* 241:233–40.

Charness, G., and M. Dufwenberg. 2006. "Promises and partnership." *Econometrica* 74:1579–1601.

Charness, G., and E. Haruvy. 2002. "Altruism, fairness and reciprocity in a gift-exchange experiment: An encompassing approach." *Games and Economic Behavior* 40:203–31.

Clutton-Brock, T. H., and G. A. Parker. 1995. "Punishment in animal societies." *Nature* 373:209–16.

Cohen, D., and I. Eshel. 1976. "On the founder effect and the evolution of altruistic traits." *Theoretical Population Biology* 10:276–302.

Colman, A. M. 1995. *Game Theory and its Applications in the Social and Biological Sciences*. Oxford: Butterworth-Heinemann.

Colman, A. M. 2006. "The puzzle of cooperation." *Nature* 440:744–45.

Cose, E. 2004. *Bone to Pick: Of Forgiveness, Reconciliation, Reparation and Revenge*. New York: Atria Books.

Cox, J. 2004. "How to identify trust and reciprocity." *Games and Economic Behavior* 46:260–81.

Cressman, R. 2003. *Evolutionary Dynamics and Extensive Form Games*. Cambridge, MA: MIT Press.

Cross, J., and M. Guyer. 1980. *Social Traps*. Ann Arbor: University of Michigan Press.

Dawes, R. 1980. "Social Dilemmas." *Annual Review of Psychology* 31:169–93.

Dawes, C. T., J. H. Fowler, T. Johnson, R. McElreath, and O. Smirnov. 2007. "Egalitarian motives in humans." *Nature* 446:794–96.

Dawkins, R. 1989. *The Selfish Gene*, 2nd ed. Oxford: Oxford University Press.

de Bruine, L. 2005. "Facial resemblance enhances trust." *Proceedings of the Royal Society B* 269:1307-12.

de Quervain, D. J.-F., U. Fischbacher, V. Treyer, M. Schellhammer, U. Schnyder, A. Buck, and E. Fehr. 2004. "The neural basis of altruistic punishment." *Science* 305:1254–58.

de Waal, F. 1996. *Good Natured*. Cambridge, MA: Harvard University Press.

De Silva, H., C. Hauert, A. Traulsen, and K. Sigmund. 2009. "Freedom, enforcement, and the social dilemma of strong altruism." *Journal of Evolutionary Economics* (forthcoming).

Dickinson, D. L. 2001. "The carrot vs. the stick in work team motivation." *Experimental Economics* 4:107–24.

Dieckmann, U., and R. Law. 1996. "The dynamical theory of co-evolution." *Journal of Mathematical Biology* 34:579–612.

Dieckmann, U., and J.A.J. Metz, eds. 2009. *Elements of Adaptive Dynamics*. Cambridge: Cambridge University Press.

Doebeli, M., C. Hauert, and T. Killingback. 2004. "The evolutionary origin of cooperators and defectors." *Science* 306:859–62.

Dufwenberg, M., U. Gneezy, W. Gueth, and E. van Damme. 2001. "Direct vs. indirect reciprocation: An experiment." *Homo Oeconomicus* 18:19–30.

Dugatkin, L. A. 1997. *Cooperation among Animals: An Evolutionary Perspective*. New York: Oxford University Press.

Dunbar, R. 1996. *Grooming, Gossip and the Evolution of Language*. Cambridge, MA: Harvard University Press.

Durrett, R., and S. A. Levin. 1994. "The importance of being discrete and spatial." *Theoretical Population Biology* 46:363–94.

Eckel, C., and R. Wilson. 2004. "Is trust a risky decision?" *Journal of Economic Behavior and Organization* 55:447–65.

Ellison, G. 1994. "Cooperation in the Prisoner's Dilemma with anonymous random matching." *Review of Economic Studies* 61:567–88.

Engelmann, D., and U. Fischbacher. 2002. "Indirect reciprocity and strategic reputation-building in an experimental helping game." Working paper, University of Zürich.

Falk, A., E. Fehr, and U. Fischbacher. 2005. "Driving forces behind informal sanctions." *Econometrica* 73:2017–30.

Fehr, E. 2004. "The neural basis of altruistic punishment." *Science* 305:1254–58.

Fehr, E., and U. Fischbacher. 2003. "The nature of human altruism." *Nature* 425: 785–91.

Fehr, E., and U. Fischbacher. 2004. "Third-party punishment and social norms." *Evolution of Human Behavior* 25:63–87.

Fehr, E., and S. Gächter. 2000. "Cooperation and punishment in public goods experiments." *American Economic Review* 90:980–94.

Fehr, E., and S. Gächter. 2002. "Altruistic punishment in humans." *Nature* 415:137–40.

Fehr, E., and J. Henrich. 2003. "Is strong reciprocity a maladaptation? On the evolutionary foundations of human altruism." In *The Genetical and Cultural Evolution of Cooperation*, ed. P. Hammerstein, 55–82. Cambridge: Cambridge University Press.

Fehr, E., and B. Rockenbach. 2003. "Detrimental effects of sanctions on human altruism." *Nature* 422:137–40.

Fehr, E., and K. Schmidt. 1999. "A theory of fairness, competition and cooperation." *Quarterly Journal of Economics* 114:817–68.

Ferrière, R. 1998. "Help and you shall be helped." *Nature* 393:517–19.

Ferrière, R., and R. Michod. 1996. "The evolution of cooperation in spatially heterogenous populations." *American Naturalist* 147:692–717.

Fishman, M. A. 2003. "Indirect reciprocity among imperfect individuals." *Journal of Theoretical Biology* 225:285–92.

Fishman, M. A., A. Lotem, and L. Stone. 2001. "Heterogeneity stabilizes reciprocal altruism interactions." *Journal of Theoretical Biology* 209:87–95.

Fletcher, J., and M. Zwick. 2004. "Strong altruism can evolve in randomly formed groups." *Journal of Theoretical Biology* 228:303–13.

Fletcher, J., and M. Zwick. 2007. "The evolution of altruism: Game theory in multi-level selection and inclusive fitness." *Journal of Theoretical Biology* 245:26–36.

Fowler, J. H. 2005a. "Human cooperation: Second-order free-riding problem solved?" *Nature* 437:E8.

Fowler, J. H. 2005b. "Altruistic punishment and the origin of cooperation." *Proceedings of the National Academy of Sciences* 10219:7047–49.

Frank, R. H. 1988. *Passion Within Reason: The Strategic Role of the Emotions*. New York: Norton.

Frank, S. A. 1998. *Foundations of Social Evolution*. Princeton: Princeton University Press.

Frank, S. A. 2003. "Repression of competition and the evolution of cooperation." *Evolution* 57:693–705.

Frean, M. R. 1994. "The Prisoner's Dilemma without synchrony." *Proceedings of the Royal Society B* 257:75–79.

Fudenberg, D., and L. Levine. 1998. *The Theory of Learning in Games*. Cambridge, MA: MIT Press.

Fudenberg, D., and E. Maskin. 1986. "The folk theorem in repeated games with discounting or with incomplete information." *Econometrica* 50:533–54.

Fudenberg, D., and E. Maskin. 1990. "Evolution and Cooperation in noisy repeated games." *American Economic Review* 80:274–79.

Gardner, A., and S. A. West. 2004. "Cooperation and punishment, especially in humans." *American Naturalist* 164:753–64.

Gaunersdorfer, A., J. Hofbauer, and K. Sigmund. 1991. "The dynamics of asymmetric games." *Theoretical Population Biology* 29:345–57.

Gintis, H. 2000. *Game Theory Evolving*. Princeton: Princeton University Press.

Gintis, H., S. Bowles, R. Boyd, and E. Fehr. 2003. "Explaining altruistic behavior in humans." *Evolution and Human Behavior* 24:153–72.

Gintis, H., S. Bowles, R. Boyd, and E. Fehr, eds. 2005. *Moral Sentiments and Material Interests: The Foundations of Cooperation in Economic Life*. Cambridge, MA: MIT Press.

Gneezy, U., and A. Rustichini. 2000. "A fine is a price." *Journal of Legal Studies* 29:1–17.

Grafen, A. 1984. "Natural selection, kin selection and group selection." In *Behavioral Ecology*, 2nd ed., ed. J. R. Krebs and N. B. Davies, 62–84. Oxford: Blackwell Scientific Publications.

Grafen, A. 2000. "Developments of the Price equation and natural selection under uncertainty." *Proceedings of the Royal Society B* 267:1223–27.

Gürerk, O., B. Irlenbusch, and B. Rockenbach. 2006. "The competitive advantage of sanctioning institutions." *Science* 312:108–11.

Güth, W., M. Knigstein, N. Marchand, and K. Nehring. 2001. "Trust and reciprocity in the investment game with indirect reward." *Homo Oeconomicus* 18:241–62.

Güth, W., R. Schmittberger, and B. Schwarze. 1982. "An experimental analysis of ultimatum bargaining." *Journal of Economic Behavior and Organization* 3:367–88.

Haidt, J. 2007. "The new synthesis in moral psychology." *Science* 316:998–1002.

Haley, K. and D. Fessler. 2005. "Nobody's watching? Subtle cues affect generosity in an anonymous economic game." *Evolution of Human Behavior* 26:245–56.

Hamilton, W. D. 1996. *Narrow Roads of Geneland: Collected Papers I*. New York: Freeman.

Hammerstein, P., ed. 2003. *Genetic and Cultural Evolution of Cooperation*. Cambridge, MA: MIT Press.

Harbaugh, W. T. 1998. "The prestige motive for making charitable transfers." *American Economic Review* 88:277–89.

Hardin, G. 1968. "The tragedy of the commons." *Science* 162:1243–48.

Härdling, R. 2007. "Fairness evolution in the ultimatum game is a function of reward size." *Journal of Theoretical Biology* 246:720–24.

Hauert, C., S. De Monte, J. Hofbauer, and K. Sigmund. 2002a. "Volunteering as a Red Queen mechanism for cooperation." *Science* 296:1129–32.

Hauert, C., S. De Monte, J. Hofbauer, and K. Sigmund. 2002b. "Replicator dynamics for optional public goods games." *Journal of Theoretical Biology* 218:187–94.

Hauert, C., and M. Doebeli. 2004. "Spatial structure often inhibits the evolution of cooperation in the snowdrift game." *Nature* 428:643–46.

Hauert, C., N. Haiden, and K. Sigmund. 2004. "The dynamics of public goods." *Discrete and Continuous Dynamical Systems B* 4:575–85.

Hauert, C., and G. Szabo. 2003. "Prisoner's dilemma and public goods games in different geometries: Compulsory versus voluntary interactions." *Complexity* 8:31–38.

Hauert, C., A. Traulsen, H. Brandt, M. A. Nowak, and K. Sigmund. 2007. "Between freedom and coercion: the emergence of altruistic punishment." *Science* 316:1905–7.

Hauser, M. 2006. *Moral Minds*. New York: Harper Collins.

Heckathorn, D. D. 1996. "The dynamics and dilemmas of collective action." *American Sociological Review* 61:250–77.

Helbing, D. 1992. "A mathematical model for behavioural changes by pair interactions." In *Economic Evolution and Demographic Change: Formal Models in Social Sciences*, ed. G. Haag, U. Mueller, and K. G. Troitzsch, 330–48. Berlin: Springer.

Henrich, J. 2006. "Costly punishment across human societies." *Science* 312:176–77.

Henrich, J. and R. Boyd. 2001. "Why people punish defectors." *Journal of Theoretical Biology* 208:79–89.

Herrmann, B., U. Thöni, and S. Gächter. 2008. "Antisocial punishment across societies." *Science* 319:1362–67.

Hilbe, C. 2008. "Memory-one Strategien für iterierte Spiele mit Fehlern." Diploma thesis, Vienna.

Hofbauer, J. 2000. "From Nash and Brown to Maynard Smith: Equilibria, dynamics and ESS." *Selection* 1:81–88.

Hofbauer, J., P. Schuster, and K. Sigmund. 1979. "A note on evolutionarily stable strategies and game dynamics." *Journal of Theoretical Biology* 81:609–12.

Hofbauer, J., and K. Sigmund. 1990. "Adaptive dynamics and evolutionary stability." *Applied Mathematics Letters* 3:75–79.

Hofbauer, J., and K. Sigmund. 1998. *Evolutionary Games and Population Dynamics*, Cambridge: Cambridge University Press.

Hofbauer, J., and K. Sigmund. 2003. "Evolutionary game dynamics." *Bulletin of the American Mathematical Society* 40:479–519.

Hutson, V.C.L., and G. T. Vickers. 1995. "The spatial struggle of tit-for-tat and defect." *Philosophical Transactions of the Royal Society B* 348:393–404.

Imhof, L., D. Fudenberg, and M. A. Nowak. 2007. "Tit-for-tat or win-stay, lose-shift?" *Journal of Theoretical Biology* 247:574–80.

Isaac, R. M., and J. M. Walker. 1988. "Communication and free-riding behaviour: The voluntary contribution mechanism." *Economic Inquiry* 26:585–608.

Johnson, D., and J. Bering. 2006. "Hand of God, mind of man: Punishment and cognition in the evolution of cooperation." *Evolutionary Psychology* 4:219–33.

Johnson, D., P. Stopka, and J. Bell. 2002. "Individual variation evades the Prisoner's Dilemma." *BMC Evolutionary Biology* 2:1–8.

Kagel, J., and A. Roth. 1995. *The Handbook of Experimental Economics*. Princeton: Princeton University Press.

Kandori, M. 1992. "Social norms and community enforcement." *The Review of Economic Studies* 59:63–80.

Kandori, M., G. J. Mailath, and R. Rob. 1993. "Learning, mutation, and long run equilibria in games." *Econometrica* 61:29–56.

Kendall, G., X. Yao, and S.Y. Chong, eds. 2007. *The Iterated Prisoner's Dilemma: 20 Years On*. Singapore: World Scientific.

Keser, C. 2003. "Experimental games for the design of reputation management systems." *IBM Systems Journal* 42:498–506.

Kiers, E., R. A. Rousseau, S. A. West, and R. F. Denison. 2003. "Host sanctions and the legume-rhizobium mutualism." *Nature* 425:78–81.

Killingback, T., J. Bieri, and T. Flatt. 2006. "Evolution in group-structured populations can solve the tragedy of the commons." *Proceedings of the Royal Society B* 273:1477–81.

Killingback, T., and M. Doebeli. 2002. "The continuous Prisoner's Dilemma and the evolution of cooperation through reciprocal altruism with variable investments." *American Naturalist* 160:421–38.

Kiyonari, T., P. Barclay, M. Wilson, and M. Daly. 2004. "Second-order punishment in one-shot social dilemma." *International Journal of Psychology* 39:329.

Kollock, P. 1993. "An eye for an eye leaves everyone blind: Cooperation and accounting systems." *American Sociological Review* 58:768–86.

Kollock, P. 1998. "Social dilemmas: The anatomy of cooperation." *Annual Review of Sociology* 24:183–214.

Kraines, D., and V. Kraines. 1989. "Pavlov and the Prisoner's Dilemma." *Theory and Decision* 26:47–79.

Kreps, D. M., and R. Wilson. 1982. "Reputation and imperfect information." *Journal of Economic Theory* 27:253–79.

Kurzban, R., P. DeScioli, and E. O'Brien. 2007. "Audience effects on moralistic punishment." *Evolution of Human Behavior* 28:75–84.

Kurzban, R., and D. Houser. 2005. "An experimental investigation of cooperative types in human groups: A complement to evolutionary theory and simulations." *Proceedings of the National Academy of Sciences* 102:1803–7.

Leimar, O. 1997. "Repeated games: A state space approach." *Journal of Theoretical Biology* 184:471–98.

Leimar, O., and P. Hammerstein. 2001. "Evolution of cooperation through indirect reciprocation." *Proceedings of the Royal Society B* 268:745–53.

Lehmann, L., and L. Keller. 2006. "The evolution of cooperation and altruism—a general framework and a classification of models." *Journal of Evolutionary Biology* 19:1365–76.

Lehmann, L., F. Rousset, D. Roze, and L. Keller. 2007. "Strong reciprocity or strong ferocity? A population genetic view of the evolution of altruistic punishment." *American Naturalist* 170:21–36.

Levin, S. A. 1999. *Fragile Dominion: Complexity and the Commons*. Reading, MA: Perseus Books.

Liebermann, E., C. Hauert, and M. A. Nowak. 2005. "Evolutionary dynamics on graphs." *Nature* 433:312–16.

Lindgren, K. 1991. "Evolutionary phenomena in simple dynamics." In *Artificial Life II*, Santa Fe Institute for Studies in the Sciences of Complexity, vol. 10, ed. C. G. Langton, et al., 295–312. Boulder, CO: Westview Press.

Lindgren, K., and M. G. Nordahl. 1994. "Evolutionary dynamics of spatial games." *Physica D* 75:292–309.

Lotem, A., M. A. Fishman, and L. Stone. 1999. "Evolution of cooperation between individuals." *Nature* 400:226–27.

Lotem, A., M. A. Fishman, and L. Stone. 2002. "Evolution of unconditional altruism through signaling benefits." *Proceedings of the Royal Society B* 270:199–205.

Luce, D., and H. Raiffa. 1957. *Games and Decisions*. New York: Wiley.

Masclet, D., C. Noussair, S. Tucker, and M. C. Villeval. 2003. "Monetary and non-monetary punishment in the voluntary contributions mechanism." *American Economic Review* 93:366–80.

Masuda, N., and H. Ohtsuki. 2007. "Tag-based indirect reciprocity by incomplete social information." *Proceedings of the Royal Society B* 274:689–95.

Matsui, A. 1992. "Best response dynamics and socially stable strategies." *Journal of Economic Theory* 57:343–62.

May, R. M. 1987. "More evolution of cooperation." *Nature* 327:15–17.

Maynard Smith, J. 1982. *Evolution and the Theory of Games*. Cambridge: Cambridge University Press.

McCabe K., M. L. Rigdon, and V. L. Smith. 2003. "Positive reciprocity and intentions in Trust games." *Journal of Economic Behavior and Organization* 52:267–75.

McNamara, J., Z. Barta, and A. Houston. 2004. "Variation in behaviour promotes cooperation in the Prisoner's Dilemma game." *Nature* 428:745–48.

Milinski, M. 1987. "Tit For Tat in sticklebacks and the evolution of cooperation." *Nature* 325:434–35.

Milinski, M., D. Semmann, T.C.M. Bakker, and H. J. Krambeck. 2001. "Cooperation through indirect reciprocity: Image scoring or standing strategy?" *Proceedings of the Royal Society B* 268:2495–2501.

Milinski, M., D. Semmann, and H. J. Krambeck. 2002a. "Donors in charity gain in both indirect reciprocity and political reputation." *Proceedings of the Royal Society B* 269:881–83.

Milinski, M., D. Semmann, and H. J. Krambeck. 2002b. "Reputation helps solve the Tragedy of the Commons." *Nature* 415:424–26.

Milinski, M., and C. Wedekind. 1998. "Working memory constrains human cooperation in the Prisoner's Dilemma." *Proceedings of the National Academy of Science* 95:13755–58.

Mohtashemi, M., and L. Mui. 2003. "Evolution of indirect reciprocity by social information: The role of trust and reputation in evolution of altruism." *Journal of Theoretical Biology* 223:523–31.

Molander, P. 1985. "The optimal level of generosity in a selfish, uncertain environment." *Journal of Conflict Resolution* 29:611–18.

Nakamaru, M., and Y. Iwasa. 2005. "The evolution of altruism by costly punishment in lattice-structured populations: Score-dependent viability vs score-dependent fertility." *Evolutionary Ecology Research* 7:853–70.

Nakamaru, M., and Y. Iwasa. 2006. "The evolution of altruism and punishment: Role of the selfish punisher." *Journal of Theoretical Biology* 240:475–88.

Neill, D. B. 2001. "Optimality under noise: Higher memory strategies for the alternating Prisoner's Dilemma." *Journal of Theoretical Biology* 211:159–80.

Nowak, M. A. 2006a. *Evolutionary Dynamics*. Cambridge, MA: Harvard University Press.

Nowak, M. A. 2006b. "Five rules for the evolution of cooperation." *Science* 314: 1560–63.

Nowak, M. A., S. Bonhoeffer, and R. M. May. 1994. "Spatial games and the maintenance of cooperation." *Proceedings of the National Academy of Sciences* 91:4877–81.

Nowak, M. A., and R. M. May. 1992. "Evolutionary games and spatial chaos." *Nature* 359:826–29.

Nowak, M. A., K. Page, and K. Sigmund. 2000. "Fairness versus reason in the Ultimatum game." *Science* 289:1773–75.

Nowak, M. A., and S. Roch. 2007. "Upstream reciprocity and the evolution of gratitude." *Proceedings of the Royal Society B* 274:605–9.

Nowak, M. A., A. Sasaki, C. Taylor, and D. Fudenberg. 2004. "Emergence of cooperation and evolutionary stability in finite populations." *Nature* 428:646–50.

Nowak, M. A., and K. Sigmund. 1989. "Game dynamical aspects of the Prisoner's Dilemma." *Applied Mathematics and Computation* 30:191–213.

Nowak, M. A., and K. Sigmund. 1990. "The evolution of stochastic strategies in the Prisoner's Dilemma." *Acta Applicandae Mathematicae* 20:247–65.

Nowak, M. A., and K. Sigmund. 1992. "Tit for Tat in heterogeneous populations." *Nature* 355:250–53.

Nowak, M. A., and K. Sigmund. 1993. "Win-stay, lose-shift outperforms tit-for-tat." *Nature* 364:56–58.

Nowak, M. A., K. Sigmund, and E. El-Sedy. 1995. "Automata, repeated games, and noise." *Journal of Mathematical Biology* 33:703–22.

Nowak, M. A., and K. Sigmund. 1998a. "Evolution of indirect reciprocity by image scoring." *Nature* 282:462–66.

Nowak, M. A., and K. Sigmund. 1998b. "The dynamics of indirect reciprocity." *Journal of Theoretical Biology* 194:561–74.

Nowak, M. A., and K. Sigmund. 1994. "The alternating Prisoner's Dilemma." *Journal of Theoretical Biology* 168:219–26.

Nowak, M. A., and K. Sigmund. 2005. "Evolution of indirect reciprocity." *Nature* 437:1291–98.

O'Gorman, R., D. S. Wilson, and R. R. Miller. 2005. "Altruistic punishing and helping differ in sensitivity to relatedness, friendship, and future interactions." *Evolution of Human Behavior* 26:375–87.

Ohtsuki, H. 2004. "Reactive strategies in indirect reciprocity." *Journal of Theoretical Biology* 227:299–314.

Ohtsuki H., and Y. Iwasa. 2004. "How should we define goodness?—Reputation dynamics in indirect reciprocity." *Journal of Theoretical Biology* 231:107–20.

Ohtsuki, H., and Y. Iwasa. 2006. "The leading eight: Social norms that can maintain cooperation by indirect reciprocity." *Journal of Theoretical Biology* 239:435–44.

Ohtsuki, H. and Y. Iwasa. 2007. "Global analyses of evolutionary dynamics and exhaustive search for social norms that maintain cooperation by reputation." *Journal of Theoretical Biology* 244:518–31.

Okuno-Fujiwara, M., and A. Postlewaite. 1995. "Social norms in matching games." *Games and Economic Behavior* 9:79–109.

Olson, M. 1965. *The Logic of Collective Action*. Cambridge, MA: Harvard University Press.

Orbell, J. H., and R. M. Dawes. 1993. "Social welfare, cooperator's advantage, and the option of not playing the game." *American Sociological Review* 58:787–800.

Ostrom, E. 1990. *Governing the Commons*. Cambridge: Cambridge University Press.

Ostrom, E., and J. Walker. 2003. *Trust and Reciprocity: Interdisciplinary Lessons from Experimental Research*. New York: Russell Sage Fundation.

Pacheco, J., F. Santos, and F. Chalub. 2006. "Stern-judging: A simple, successful norm which promotes cooperation under indirect reciprocity." *PLoS Computational Biology* 2:e178.

Pacheco, J., A. Traulsen, H. Ohtsuki, and M. A. Nowak. 2008. "Repeated games and direct reciprocity under active linking." *Journal of Theoretical Biology* 250:723–31.

Page, K. M., M. A. Nowak, and K. Sigmund. 2000. "The spatial Ultimatum game." *Proceedings of the Royal Society B* 267:2177–82.

Page, K., and M. A. Nowak. 2001. "A generalized adaptive dynamics framework can describe the evolutionary Ultimatum game." *Journal of Theoretical Biology* 209:173–79.

Page, K. M., and M. A. Nowak. 2002. "Unifying evolutionary dynamics." *Journal of Theoretical Biology* 219:93–98.

Panchanathan, K., and R. Boyd. 2003. "A tale of two defectors: The importance of standing for evolution of indirect reciprocity." *Journal of Theoretical Biology* 224:115–26.

Panchanathan, K., and R. Boyd. 2004. "Indirect reciprocity can stabilize cooperation without the second-order free rider problem." *Nature* 432:499–502.

Pepper, J. W. 2000. "Relatedness in trait group models of social evolution." *Journal of Theoretical Biology* 206:355–68.

Peyton Young, H., and D. Foster. 1995. "Learning dynamics in games with stochastic perturbations." *Games and Economic Behavior* 11:330–63.

Pfeiffer, T., C. Rutte, T. Killingback, M. Taborsky, and S. Bonhoeffer. 2005. "Evolution of cooperation by generalized reciprocity." *Proceedings of the Royal Society B* 272:1115–20.

Pflügl, C. 2007. "Evolution altruistischen Verhaltens durch Gruppenselektion." Diploma thesis, Vienna.

Pollock, G. B., and L. A. Dugatkin. 1992. "Reciprocity and the evolution of reputation." *Journal of Theoretical Biology* 159:25–37.

Poundstone, W. 1992. *Prisoner's Dilemma*. New York: Doubleday.

Price, G. R. 1970. "Selection and covariance." *Nature* 277:520–21.

Price, M. E., L. Cosmides, and J. Tooby. 2002. "Punitive sentiment as an anti-free rider psychological device." *Evolution of Human Behavior* 23:203–31.

Rankin, F. R. 2003. "Communication in Ultimatum games." *Economic Letters* 81: 267–71.

Rapoport, A., and A. Chammah. 1965. *The Prisoner's Dilemma*. Ann Arbor: University of Michigan Press.

Rege, M., and K. Telle. 2004. "The impact of social approval and framing on cooperation in public good situations." *Journal of Public Economics* 88:1625–44.

Richerson, P., and R. Boyd. 2005. *Not by Genes Alone*. Chicago: University of Chicago Press.

Ridley, M. 1997. *The Origins of Virtue*. New York: Viking Press.

Rilling, J., D. Gutman, T. Zeh, D. Pagnoni, G. Berns, and C. Kilts. 2002. "A neural basis for social cooperation." *Neuron* 35:395–495.

Roberts, G. 2008. "Evolution of direct and indirect reciprocity." *Proceedings of the Royal Society B* 275:173–79.

Roberts, G., and T. N. Sherratt. 1998. "Development of cooperative relationships through increasing investment." *Nature* 394:175–79.

Rockenbach, B., and M. Milinski. 2006. "The efficient interaction of indirect reciprocity and costly punishment." *Nature* 444:718–23.

Rosenthal, R. W. 1979. "Sequences of games with varying opponents." *Econometrica* 47:1353–66.

Rousset, F. 2004. *Genetic structure and selection in subdivided populations*. Princeton: Princeton University Press.

Rubinstein, A. 1986. "Finite automata play the repeated Prisoner's Dilemma." *Journal of Economic Theory* 39:83–96.

Rutte, C., and M. Taborsky. 2007. "Generalized reciprocity in rats." *PLoS Biology* 5(7):e196.

Sachs, J., U. G. Mueller, T. P. Wilcox, and J. J. Bull. 2004. "The evolution of cooperation." *Quarterly Review of Biology* 79:135–60.

Sandholm, W. H. 2009. *Population Games and Evolutionary Dynamics*. Harvard: MIT Press.

Sanfey, A., J. Rilling, J. Aronson, L. Nystrom, and J. Cohen. 2003. "The neural basis of economic decision making in the Ultimatum game." *Science* 300:1755–58.

Santos, F. C., J. M. Pacheco, and T. Lenaerts. 2006. "Evolutionary dynamics of social dilemmas in structured heterogeneous populations." *Proceedings of the National Academy of Science* 103:3490–94.

Schelling, T. C., 1971. "Dynamic models of segregation." *Journal of Mathematical Sociology* 1:143–86.

Schelling, T. C. 1978. *Micromotives and Macrobehavior*. New York: Norton.

Schlag, K. H. 1998. "Why imitate, and if so, how? A boundedly rational approach to multi-armed bandits." *Journal of Economic Theory* 78:130–56.

Schuster, P., and K. Sigmund. 1983. "Replicator dynamics." *Journal of Theoretical Biology* 100:533–38.

Sefton, M., R. Shupp, and J. Walker. 2007. "The effects of rewards and sanctions in provision of public goods." *Economic Inquiry* 45:671–90.

Seinen, I., and A. Schram. 2001. "Social status and group norms: Indirect reciprocity in a repeated helping experiment." *European Economic Review* 50:581–602.

Sell, J., and R. K. Wilson. 1999. "The maintenance of cooperation: Expectations of future interactions and the trigger of group punishment." *Social Forces* 77:1551–70.

Selten, R., and P. Hammerstein. 1984. "Gaps in Harley's argument on evolutionarily stable learning rules and the logic of tit for tat." *Behavior and Brain Sciences* 7:115–16.

Semmann, D., H. J. Krambeck, and M. Milinski. 2003. "Volunteering leads to rock-paper-scissors dynamics in a public goods game." *Nature* 425:390–93.

Semmann, D., H. J. Krambeck, and M. Milinski. 2004. "Strategic investment in reputation." *Journal of Behavioral Ecology and Sociobiology* 56:248–52.

Sheratt, T. N., and G. Roberts. 2001. "The importance of phenotypic defectors in stabilizing reciprocal altruism." *Behavioral Ecology* 12:313–17.

Sigmund, K. 1995. *Games of Life*. Harmondsworth: Penguin Press.

Sigmund, K. 2007. "Punish or perish? Retaliation and collaboration among humans." *Trends in Ecology and Evolution* 22:593–600.

Sigmund, K., C. Hauert, and M. A. Nowak. 2001. "Reward and punishment." *Proceedings of the National Academy of Sciences* 98:10757–62.

Silk, J. B. 2006. "Who are more helpful, humans or chimpanzees?" *Science* 311:1248–49.

Skyrms, B. 2004. *The Stag Hunt and the Evolution of Social Structure*. Cambridge: Cambridge University Press.

Skyrms, B., and R. Pemantle. 2000. "A dynamic model of social network formation." *Proceedings of the National Academy of Sciences* 97:9340–46.

Small, D. A., and G. Loewenstein. 2005. "The devil you know: The effects of identifiability on punishment." *Journal of Behavioral Decision Making* 18:311–18.

Sober, E., and D. S. Wilson. 1998. *Unto Others*. Cambridge, MA: Harvard University Press.

Sommerfeld, R., H. J. Krambeck, D. Semmann, and M. Milinski. 2007. "Gossip as an alternative for direct observation in games of indirect reciprocity." *Proceedings of the National Academy of Sciences* 104:17435–40.

Stephens, D. W., C. M. McLinn, and B. Stevens. 2002. "Discounting and reciprocity in an iterated Prisoner's Dilemma." *Science* 298:2216–18.

Stevens, J. R., F. A. Cushman, and M. D. Hauser. 2005. "Evolving the psychological mechanisms for cooperation." *Annual Review of Ecology, Evolution and Systematics* 36:499–518.

Sugden, R. 1986. *The Economics of Rights, Cooperation and Welfare*. Oxford: Basil Blackwell.
Suzuki, S., and E. Akiyama. 2007a. "Evolution of indirect reciprocity in groups of various sizes and comparison with direct reciprocity." *Journal of Theoretical Biology* 245:539–52.
Suzuki, S., and E. Akiyama. 2007b. "Three-person game facilitates indirect reciprocity under image scoring." *Journal of Theoretical Biology* 249:93–100.
Szabó, G., and G. Fáth. 2007. "Evolutionary games on graphs." *Physics Reports* 446:97–216.
Szabó, G., and C. Hauert. 2002. "Phase transitions and volunteering in spatial public goods games." *Physical Review Letters* 89 (118101): 1–4.
Szabó, G., and J. Vukov. 2004. "Cooperation for volunteering and partially random partnerships." *Physical Review* E 69 (036107): 1–7.
Takahashi, N., and R. Mashima. 2004. "The importance of indirect reciprocity: Is the standing strategy the answer?" Working paper, Hokkaido University.
Takahashi, N. and R. Mashima. 2006. "The importance of subjectivity in perceptual errors on the emergence of indirect reciprocity." *Journal of Theoretical Biology* 243:418–36.
Taylor, C., D. Fudenberg, A. Sasaki, and M. A. Nowak. 2004. "Evolutionary game dynamics in finite populations." *Bulletin of Mathematical Biology* 66:1621–44.
Taylor, P. D., T. Day, and G. Wild. 2007. "Evolution of cooperation in a finite homogeneous graph." *Nature* 447:469–72.
Taylor, P. D., and S. A. Frank. 1996. "How to make a kin selection model." *Journal of Theoretical Biology* 180:27–37.
Taylor, P. D., and L. Jonker. 1978. "Evolutionarily stable strategies and game dynamics." *Mathematical Biosciences* 40:145–56.
Traulsen, A., C. Hauert, H. De Silva, M. A. Nowak, and K. Sigmund. 2009. "Exploration dynamics in evolutionary games." *Proceedings of the National Academy of Sciences* 106:706–9.
Trivers, R. 1971. "The evolution of reciprocal altruism." *Quarterly Review of Biology* 46:35–57.
Trivers, R. 1985. *Social Evolution*. Menlo Park, CA: Benjamin Cummings.
Trivers, R. 2002. *Natural Selection and Social Theory: Selected Papers of Robert Trivers*. Oxford: Oxford University Press.
Trivers, R. 2006. "Reciprocal altruism: 30 years later." In *Cooperation in Primates and Humans: Mechanisms and Evolution*, ed. P. M. Kappeller and C. P. van Schaik, 67–83. Berlin: Springer Verlag.
Vincent, T., and J. S. Brown. 2005. *Evolutionary Game Theory, Natural Selection, and Darwinian Dynamics*. Cambridge: Cambridge University Press.
Wahl, L. M., and M. A. Nowak. 1999. "The continuous Prisoner's Dilemma: I. Linear reactive strategies." *Journal of Theoretical Biology* 200:307–21.
Walker, J. M., and M. Halloran. 2004. "Rewards and sanctions and the provision of public goods in one-shot settings." *Experimental Economics* 7:235–47.
Warneken, F., and M. Tomasello. 2006. "Altruistic helping in human infants and young chimpanzees." *Science* 31:1301–3.
Wedekind, C., and V. A. Braithwaite. 2002. "The long-term benefits of human generosity in indirect reciprocity." *Current Biology* 12:1012–15.

Wedekind, C., and M. Milinski. 2000. "Cooperation through image scoring in humans." *Science* 288:850–52.

Weibull, J. 1995. *Evolutionary Game Dynamics*. Cambridge, MA: MIT Press.

Wenseleers, T. and F.L.W. Ratnieks. 2006. "Comparative analysis of worker reproduction and policing in eusocial Hymenoptera supports relatedness theory." *American Naturalist* 168: E163–E179.

West, S. A., A. S. Griffin, and A. Gardner. 2007. "Social semantics: Altruism, cooperation, mutualism and strong reciprocity." *Journal of Evolutionary Biology* 20:415–32.

West, S. A., I. Pen, and A. S. Griffin. 2002. "Cooperation and competition between relatives." *Science* 296:72–75.

Xiao, E., and D. Houser. 2005. "Emotion expression in human punishment behavior." *Proceedings of the National Academy of Sciences* 102:7398–7401.

Yamagishi, T. 1986. "The provision of a sanctioning system as a public good." *Journal of Personality and Social Psychology* 51:110–16.

Yamagishi, T., N. Jin, and T. Kiyonari. 1999. "Bounded generalized reciprocity: Ingroup boasting and ingroup favoritism." *Advances in Group Processes* 16:161–97.

Zeeman, E. C. 1980. "Population dynamics from game theory." In *Global Theory of Dynamical Systems*, Lecture Notes in Mathematics 819 ed. Z. Nitecki and C. Robinson. New York: Springer.

Index